Dagmar Henrich

Influence of Material and Geometry on the Performance of Superconducting Nanowire Single-Photon Detectors

Karlsruher Schriftenreihe zur Supraleitung   Band 010

HERAUSGEBER

Prof. Dr.-Ing. M. Noe
Prof. Dr. rer. nat. M. Siegel

Eine Übersicht über alle bisher in dieser Schriftenreihe
erschienene Bände finden Sie am Ende des Buchs.

Influence of Material and Geometry on the Performance of Superconducting Nanowire Single-Photon Detectors

by
Dagmar Henrich

Dissertation, Karlsruher Institut für Technologie (KIT)
Fakultät für Elektrotechnik und Informationstechnik, 2013
Hauptreferent: Prof. Dr. Michael Siegel
Korreferent: Prof. Dr. Uli Lemmer

Impressum

 Scientific
Publishing

Karlsruher Institut für Technologie (KIT)
KIT Scientific Publishing
Straße am Forum 2
D-76131 Karlsruhe

KIT Scientific Publishing is a registered trademark of Karlsruhe
Institute of Technology. Reprint using the book cover is not allowed.

www.ksp.kit.edu

Print on Demand 2013

ISSN 1869-1765
ISBN 978-3-7315-0092-6

Influence of Material and Geometry on the Performance of Superconducting Nanowire Single-Photon Detectors

zur Erlangung des akademischen Grades eines

DOKTOR-INGENIEURS

von der Fakultät für
Elektrotechnik und Informationstechnik
des Karlsruher Instituts für Technologie (KIT)

genehmigte

DISSERTATION

von

Dipl.-phys. Dagmar Henrich

geb. in Sigmaringen

Tag der mündlichen Prüfung: 23. Juli 2013

Hauptreferent: Prof. Dr. Michael Siegel

Korreferent: Prof. Dr. Uli Lemmer

Preface

The work presented in this dissertation was conducted during my time at the *Institut für Mikro- und Nanoelektronische Systeme* and at the *Lichttechnisches Institut*. Both institutes were formerly part of the *Universität Karlsruhe (TH)* and are now part of the *Karlsruhe Institute for Technology*. I want to thank the heads of both institutes, Prof. Siegel and Prof. Lemmer, respectively, for giving me the opportunity to work on this exciting topic on the borderline between Physics and Electrical Engineering.

I especially thank my supervisor Dr. Konstantin Ilin who has given the work direction and shape. He introduced me to the world of nanotechnology and was always available for scientific discussions. I very much appreciated his prompt, extensive and helpful feedback. To my colleagues in our common office, Dr. Stefan Wünsch, Matthias Hofherr, Gerd Hammer and Matthias Arndt I am very grateful for the scientific exchange and their help on countless occasions, but even more so for the brakes in between and around the work hours. I very much enjoyed working with them and learned much about science as well as life in general. This gratitude extends to the fellow PhD students at both institutes who are too numerous to list. I share many fond memories of our time at the KIT, at conferences and workshops and owe them many ideas and tips that influenced my work. Much of the work was made possible only by the technical staff of the IMS, who fabricated the required parts in the various experimental setups. Them, and in the same way the administrative staff, I thank for their patience and diligence. Also, an indispensable part of the work was supplied by many Bachelor and Master students who have worked on sub-topics during their thesis, as well as student workers. I especially want to thank Steffen Dörner, Yannick Luck, Laura Rehm, Patrick Reichensperger, Julia Toussaint and Klevis Ylli for their valuable work and the great time.

Furthermore I thank the graduate school *Karsruhe School of Optics and Photonics*, not least for the financial support, but also for offering many social and educational opportunities. I have met many graduate students from institutes all across the university and the fruitful discussions often resulted in helpful suggestions or exchange of equipment. My final thanks go to my loving family and my husband Christian for their unfaltering emotional support and motivation throughout the whole time I was working at this thesis.

Karlsruhe,

im Juli 2013

Dagmar Henrich

Karlsruher Institut für Technologie (KIT)

Kurzfassung

Diese Arbeit untersucht den Einfluss von material- und geometrieabhängigen Faktoren auf die zentralen Kenngrößen von supraleitenden Einzelphotonendetektoren (Superconducting Nanowire Single-Photon Detector, SNSPD). SNSPD basieren auf einem Supraleiter, der als Nanodraht strukturiert ist und von einem Strom knapp unterhalb der kritischen Stromstärke durchflossen wird. Die Absorption eines Photons in dem Supraleiter verursacht den lokalen Zusammenbruch der Supraleitung, wodurch eine messbare Spannung abfällt. Die Möglichkeit, einzelne Photonen zu detektieren, ist dabei nicht nur für den Einsatz bei sehr geringen Lichtintensitäten interessant, sondern für manche Anwendungen wie Quantenexperimente zwingend notwendig.

Im optischen Spektralbereich wird die Einzelphotonendetektion bislang dominiert von Halbleiterdetektoren, die jedoch im infraroten Spektrum aufgrund ihrer Bandlücke an eine fundamentale Grenze stoßen. Gerade hier stellen SNSPD eine konkurrenzfähige Alternative dar und weisen zusätzlich eine geringe Dunkelzählrate auf, können aufgrund schneller Relaxationszeiten hohe Zählraten erzielen und für Messungen mit hoher Zeitauflösung eingesetzt werden. Allerdings weist die Detektionseffizienz oberhalb einer gewissen Wellenlänge einen exponentiellen Abfall auf, der die Leistung im infraroten Bereich stark beeinträchtigt. Ziel dieser Arbeit ist es, die Zusammenhänge zwischen Abfall der Detektionseffizienz und den Materialparametern sowie der Detektorgeometrie qualitativ und quantitativ zu untersuchen, und so die Detektionseffizienz von SNSPD im infraroten Spektralbereich zu erhöhen. Gleichzeitig soll dabei eine möglichst geringe Dunkelzählrate und eine hohe maximale Zählrate erhalten bleiben.

Um den Einfluss der Materialparameter auf den SNSPD zu bestimmen, werden zunächst NbN Dünnschichten untersucht. NbN ist ein weit verbreitetes Material für den Einsatz als SNSPD und kann durch Verschiebung des stöchiometrischen Verhältnisses von Niob zu Stickstoff fast stufenlos verändert werden. Dies geschieht durch die Variation der Depositionsparameter bei der reaktiven Abscheidung der Schicht. Neben der detaillierten Charakterisierung der supraleitenden und resistiven Eigenschaften der NbN Dünnschichten werden die Filme auch mittels THz-Spektroskopie untersucht. Dies erlaubt eine direkte Messung der supraleitenden Energielücke im Temperaturbereich von T_C bis 4,2 K, bei der SNSPD betrieben werden. Desweiteren werden ellipsometrische Messungen in einem breiten spektralen Bereich durchgeführt und mittels eines eigens entwickelten Schichtsystems für die NbN Multilagenstruktur die optischen Eigenschaften extrahiert. Das Ergebnis ist eine umfassende Analyse der Abhängigkeit der wesentlichen Größen, die das Detektorverhalten beeinflussen, von der Materialvariation.

Die maximale Zählrate von SNSPD wird begrenzt durch die Relaxationszeit des Detektors nach dem Eintritt einem Zählereignisses. Nach Absorption eines Photons wird dessen Energie kaskadenartig an die entstandenen Quasiteilchen und darauffolgend an die Phononen weitergegeben, die schließlich aus dem Supraleiter diffundieren. Zur Untersuchung der involvierten Zeitkonstanten wurde ein Frequenzmesssystem aufgebaut, bei dem durch Überlagerung zweier Laserstrahlen eine kontinuierlich verstellbare Modulationsfrequenz erzeug wird. Die Auswertung der Messungen an NbN Dünnschichten erfolgt auf Basis des 2-Temperaturen-Modells und erlaubt die Analyse der Relaxationszeitkonstanten. Daraus ergib sich, dass sich mit der Materialzusammensetzung im Wesentlichen die Interaktionszeit zwischen Elektronen und Phononen ändert, während die Ausdiffusion der Phononen aus dem Supraleiter von der Filmdicke bestimmt wird. Minimale Relaxationszeit werden erreicht für dünne Filme und NbN mit erhöhtem Stickstoffanteil. Für typische Filmdicken in der Anwendung in SNSPD werden unabhängig von der Materialzusammensetzung ausreichende Zählraten erzielt.

Ein weiterer wesentlicher Faktor für das Erreichen hoher Detektionseffizienzen ist die Maximierung des Verhältnisses der angelegten Stromstärke zur paarbrechenden kritischen Stromstärke. Die Dimensionen eines Nanodrahts liegen im Bereich der charakteristischen Dimensionen von magnetischen Flussschläuchen, deren Eindringen in den Supraleiter die kritische Stromdichte erheblich reduzieren kann. Der Einfluss von Dicke und Breite von supraleitenden Brücken wird anhand von mehreren verschiedenen supraleitenden Materialien untersucht. Dabei werden auch zum ersten Mal Stromdichtemessungen an strukturierten $Ba(Fe,Co)_2As_2$-Filmen durchgeführt. Die Untersuchungen ergeben, dass für Temperaturen nahe T_C keine Flusschläuche in den Supraleiter eindringen und der kritische Strom nur durch die Bindungsenergie der Cooper-Paare begrenzt wird. Es wird diskutiert, wie dies durch Verringerung der Breite und geeignete Wahl des supraleitenden Materials auch bei tieferen Temperaturen erreicht werden kann. Entscheidend ist hierbei eine möglichst hohe energetische Barriere für das Eindringen der magnetischen Flussschläuche in den Supraleiter.

Diese energetische Barriere wird zusätzlich von der Geometrie des Nanodrahts beeinflusst. Bei einem Knick kommt es zu einer Stromdichteerhöhung entlang der inneren Kante und damit zu einer lokalen Absenkung der Eindringbarriere. In dem für SNSPD weit verbreiteten mäanderförmigen Design führen die abrupte Kehrtwenden an den Seiten des Detektors zu einer Reduktion der kritischen Stromstärke. Dieser Effekt wird zunächst systematisch an Nb und NbN Teststrukturen untersucht und mit theoretischen Modellen verglichen. Zudem wird die theoretische Vorhersage der effektiven Erhöhung der kritischen Stromdichte durch Anlegen eines magnetischen Feldes in geeigneten Proben experimentell verifiziert.

Basierend auf den geschilderten Untersuchungen werden im letzten Teil der Arbeit SNSPD mit verbesserten Eigenschaften vorgeschlagen und hergestellt. Zur direkten experimentellen Bestimmung der Detektoreigenschaften wurde ein fasergekoppelter kryogener Messaufbau entwickelt. Im ersten Ansatz werden SNSPD aus NbN-Schichten hergestellt, bei denen das stöchiometrische Verhältnis zwischen Niob und Sickstoff variiert wird. Die Proben mit reduziertem Stickstoffanteil weisen dabei eine deutlich verbesserte Detektionseffizienz im infraroten Bereich und gleichzeitig eine niedrigere Dunkelzählrate auf. Die Ergebnisse werden quantitativ mit den Vorhersagen des 'Hot-Spot' Modells verglichen und Diskrepanzen diskutiert. Im zweiten Ansatz wird eine alternative spiralförmige Detektorgeometrie eingesetzt, die eine verbesserte Stromdichteverteilung aufweist. Obwohl die absolute Stromdichte gegenüber dem Mäanderdesign nur schwach erhöht ist, führt die homogene Stromdichteverteilung zu einer deutlich erhöhten Detektionseffizienz. Die runde Form weist außerdem eine geringere Polarisationsabhängigkeit auf und erlaubt eine einfachere Ankopplung an optische Systeme.

Contents

List of Figures   151

List of Tables   155

List of Equations   157

Nomenclature   159

Bibliography   165

1 Introduction

The ability to detect light on a single-photon level is generally desirable for all applications that need to detect low light intensity. In scientific projects like ground-based astronomy or spectroscopy, radiation intensities are often at the quantum limit. But for certain applications, detectors with the ability to count photons are an enabling technology. In the wide field of quantum information science, the demand on detector performance that goes beyond conventional single-photon counting has increased significantly over the past years [1]. This includes applications like quantum key distribution and optical quantum computing, but also many experimental investigations on correlated photons, quantum dots and quantum emitters. In industrial applications, single-photon detectors are used in quality testing like integrated circuit evaluation.

Single-photon detection in the optical spectrum has been dominated by photomultiplier tubes (PMT) and more recently by solid state single-photon avalanche photodiodes (SPAD) [2]. Although they benefit from the fact that only moderate cooling is required (typical devices are operated well above $200\,\mathrm{K}$), the spectrum of these devices is limited by the band gap of the semiconductor that is employed, which is usually Si up to $\approx 1.1\,\mu\mathrm{m}$ (>60% detection efficiency at $550\,\mathrm{nm}$) or for longer wavelengths InGaAs up to $\approx 2.6\,\mu\mathrm{m}$ (40-50% detection efficiency at $550\,\mathrm{nm}$)[3]. Also, those devices suffer from relatively high dark count rates and afterpulsing, which requires sophisticated readout electronics to achieve top-end perfomance values. Detailed comparison of key parameters of contemporary photon-counting technologies can be found in [4] and [5].

As an alternative, superconducting nanowire single-photon detectors (SNSPD) show very promising properties for a wide variety of applications. SNSPDs offer a high counting efficiency, low dark count rates and fast time resolution in a wide spectral range. The detectors operate in many cases above the boiling point of liquid Helium and are thus applicable for the use of closed-cycle refrigerators. SNSPDs show single-photon sensitivity up to $5\,\mu\mathrm{m}$, and at telecommunications wavelengths detection efficiencies around 50% have been achieved [6]. This value can be increased further by the application of improved radiation coupling techniques to overcome the low absorption probability of the thin nanowire, which has recently been the subject of many efforts. While research on this topic is still ongoing, it has already reached a position where >90% of the photons in a fiber or waveguide can be successfully guided to the SNSPD and absorbed in the superconductor. Respectively, high detection efficiencies were demonstrated with these techniques in the optical range.

Due to their very low timing jitter, SNSPDs are also especially attractive for time-correlated single-photon counting (TCSPC) techniques in the infrared range. If the arrival time of a photon can be determined very well, which is possible with very high accuracy with modern optical sources, the jitter of the photon counter is the limiting factor of TCSPC resolution. A timing jitter of 29 ps has recently been reported for small area single and multipixel SNSPD [7].

Originally, SNSPD were used to detect optical and infrared photons, but they can principally be operated in a wide range of frequencies, for example in the x-ray spectrum [8]. Additionally, they are not limited to the detection of electromagnetic waves, but can also be used to detect particles. The particles do not need to have an electric or magnetic charge, only a kinetic energy that is high enough to cause a count event upon absorption. SNSPDs have been employed for the detection of electrons [9], neutrons, atoms and even molecules [10, 11].

However, the main field of competition for superconducting nanowire single-photon detectors is the infrared region of the electromagnetic spectrum, where detectors based on semiconductors reach their physical limitations. The band gap of a superconductor is three orders of magnitude smaller, making them promosing candidates for detection of photons with much lower energies. But to induce a count event in an SNSPD, a certain density of quasiparticles has to be generated inside the nanowire. The detection efficiency of an SNSPD reaches high values for photon energies that create a sufficient amount of quasiparticles, but reduces exponentially as the photon wavelength is increase above a certain threshold value. This intrinsic property significanlty limits the detection efficiency of the devices in the infrared range. The central motivation for the research presented in this work is to increase the detection efficiency in the infrared range by increase of this intrinsic threshold wavelength. At the same time, the two other main advantages of SNSPD, which are the low dark count rate and the high maximum count rates, should be retained within reasonable values.

The quasiparticle density which is neccessary to create a count event and the involved time constants are governed by three major influences: The material properties of the superconducting film, the relaxation processes that take place in the superconductor after the photon absorption and the transport properties of the device. These form the three main approaches that are pursued in this work. A more detailed description of the count event generation will be derived in the first chapter, where the theoretical background and the current state of SNSPD research are presented. The approaches to improve SNSPD performance are formulated more preciesly at the end of chapter 2, once the neccessary background and terminology have been introduced.

The first task is to improve the material properties of the superconducting thin film. NbN is deposited by reactive magnetron sputtering into thin films with thicknesses down to 4 nm. The material is commonly used by many groups as a basis for SNSPD devices. The stoichiometry of the NbN films is changed by the variation of the deposition conditions. The resulting influence on the material parameters relevant for SNSPD operation is investigated by cryogenic resistive and magnetic measurements, THz spectroscopy and ellipsometry (chapter 3).

The energy relaxation processes inside a NbN thin film are studied with a frequency-domain measurement technique. An experimental setup is developed, where the superconducting film is excited by modulated laser light. The response signal can be determined in a range of modulation frequency from 10 MHz up to 10 GHz, which allows conclusions on the relaxation time constants (chapter 4).

The current transport properties are dependent on the nanowire dimensions and layout. To identify the processes responsible for the generation of the critical state in a nanowire, superconducting bridges with different dimensions are investigated. Results are obtained and compared for structures made from Nb, NbN, TaN and $Ba(Fe,Co)_2As_2$. The reduction of the critical current by the geometrical layout is then studied systematically on a set of samples structured with single discontinuities (chapter 5).

Based on the results of this investigatons, SNSPD with improved properties are designed and structured from NbN thin films. The detectors with linewidths of ≈ 100 nm are patterned by electron-beam lithography and an optimized reactive ion-etching process. A fiber-based cryogenic setup is installed for the direct measurement of the detector response in a broad spectral range of $400 - 1700$ nm. The spectral detection efficiency and dark count rate of detectors with different stoichiometry and layout geometries are studied to identify the main influences on the detector performance. An improved infrared detection efficiency will be demonstrated on optimized SNSPD devices (chapter 6).

2 Superconducting nanowire single-photon detectors: theory and state of the art

For a more precise formulation of the tasks pursued in this work, a brief overview of the SNSPD theoretical models and the current state of the research is given. The mechanism of single-photon detection with a superconducting nanowire is described and the figures of merit of such detectors are defined. A recollection of previous work on the topic is given with a focus on studies of the intrinsic detection efficiency, which will be identified as the main limiting property for infrared detection. The hot-spot model is introduced in section 2.2 to describe the spectral dependence of the intrinsic detection efficiency. It provides the theoretical background for the discussion and for the analysis of results later on. On this basis, the specific approaches to improve SNSPD performance studied in this work are formulated in section 2.3.

2.1 Superconducting nanowire single-photon detectors

2.1.1 Principle of operation and figures of merit

The concept of a Superconducting Nanowire Single-Photon Detectors (SNSPD) was first demonstrated on a niobium nitride (NbN) nanowire by Gol'tsman et al. [12] and will be briefly reviewed in the following: The devices are usually made from very thin superconducting films, typically below 5 nm in thickness. The film is patterned into a line with typical widths $w \leq 100\,\text{nm}$, which is referred to as nanowire due to the fact that its cross-section is in both dimensions in the nanometer regime.

The principle of operation is schematically illustrated in Fig. 2.1. The nanowire (gray) is cooled down to temperatures well below the superconducting transition $T \ll T_C$, where the superconductor exhibits a clear discontinuous step in its current-voltage characteristic once the critical current I_C is surpassed. I_C is defined by the maximum kinetic energy of the Cooper pairs carrying the supercurrent that does not exceed the binding energy. The wire is biased with a current I_B (indicated by blue arrows) that is close to, but below, I_C. In this initial state the voltage measured across the device is zero (left graph).

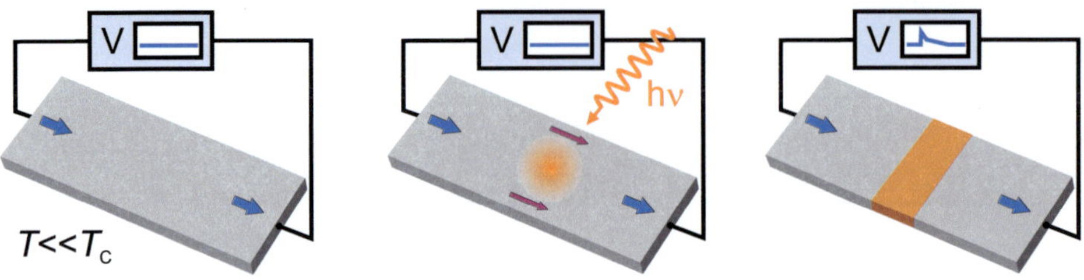

Figure 2.1: Schematic illustration of the SNSPD principle of operation: In the initial state (left), the nanowire is in the superconducting state and biased with a current (blue arrows) close to the critical current. When a photon is incident on the nanowire (middle), it may be absorbed and cause a local zone of suppressed superconductivity. This zone grows into a normal conducting region across the wire (right), that leads to a measurable voltage pulse in the readout. Afterwards, the nanowire relaxes back to its initial state.

If a single photon is incident on the wire, there is a chance that it will be absorbed by one of the Cooper pairs (middle graph). The energy of a photon is usually much larger than the superconducting energy gap Δ, for example optical photons have $E_{ph} \approx 1\,\mathrm{eV}$ and typical superconductor energy gaps are $\Delta \approx 1\,\mathrm{meV}$. This means that the pair will break into quasiparticles of high energy that start a cascade of further quasiparticle generation. The reduction of the Cooper pair density leads to region of suppressed superconductivity around the absorption site (yellow). The higher the initial photon energy E_{ph}, the higher the number of quasiparticles that are generated by this avalanche process. On the other hand, in a superconductor with a smaller energy gap Δ, less energy is needed to create a sufficient amount of quasiparticles, making it in principle more sensitive for longer wavelengths.

The supercurrent through the cross-section of the absorption site now has to be sustained by a reduced amount of Cooper pairs. This effectively causes a decrease in I_C, also in the regions that were not directly affected (purple arrows). Since I_B was already close to I_C, the critical current is then exceeded in the whole cross-sectional area of the nanowire, which transits to the normalconducting state (yellow region in right graph). The bias current now faces a normal conducting zone in its path, giving rise to an abrupt voltage pulse which is registered at the readout as a count event.

The region of reduced Cooper pair density caused by the photon absorption has to fulfill two conditions to result into a normalconducting zone:

- The processes of the quasiparticle avalanche must have time scales shorter than the relaxation channels, e.g. out-diffusion or energy transfer to the lattice, so that the reduction of the Cooper pair density remains local.

- The reduction has to be significant compared to the number of states in the considered volume. This number is given on one hand by the density of states. Materials with a low density of states are preferable, because a smaller number of Cooper pairs have to be broken to create the normalconducting zone. On the other hand, the volume has to be sufficiently small, specifically the cross-section of the wire. This is the main reason for the dimensions of the wire in the nanometer regime. The smaller the cross-section the less photon energy is required to create the normalconducting zone that causes the count event.

Once the energy deposited by the photon dissipates from of the normal zone, and the cooling power is sufficient to counter the heat dissipation in the resistive region by the bias current, the nanowire will eventually fall back to its initial superconducting state. The time constant for this process depends foremost on the inductance of the device and the electrical circuitry. For typical SNSPD, the effective device dead time is in the range of nanoseconds, but can be improved much below this value if required.

The most important property of a SNSPD is the probability with which an incident photon leads to a count event, the detection efficiency. The exact definition of this parameter often varies in literature, depending on the method of determination, the exact setup and how much of the coupling losses are included in the definition. Generally speaking, any setup has a system detection efficiency (SDE), which is the number of count events registered by the system divided by the number of photons that were input to the system during a given time interval. The overall SDE can be split up into three parts:

- The optical coupling efficiency (OCE) describes the efficiency of guiding the photons to the detector area. It is a property of the system setup and the options that can be employed to improve it depend strongly on the application considered.

- The absorption probability (ABS) gives the fraction of photons incident on the detector area that are actually absorbed inside the nanowire. It depends mainly on the coupling of the photon to the wire as well as its optical path length inside the superconductor and the material's absorption coefficient.

- The intrinsic detection efficiency (IDE) of the device itself, that is, the number of output pulses the detector delivers divided by the number of photons that have been absorbed inside the nanowire. The product of IDE and ABS is often referred to as detection efficiency (DE) in the literature.

The efficiency of the readout, i.e. the fraction of output pulses delivered by the detector that are actually registered as count events by the setup, is for normal circumstances approximately 100% and is usually neglected in the considerations. All of these contributions to the detection efficiency have a spectral dependence which together define the spectral bandwidth of the system. The extend to which this bandwidth reaches into the long-wavelength range is an important figure of merit for an SNSPD and is often described with the so-called cut-off wavelength. It will be described in more detail in section 2.2.

There is a certain probability that count events occur even though no photons are incident on the detector. This is called the dark count rate (*DCR*) and it can be understood as the equivalent of noise for such a detection principle. These 'false' counts are a limit for the minimum photon flux that can be well resolved by the detector. Note that the *DCR* is not connected to the the noise of the readout channel.

Depending on the application, additional figures of merit concern the speed of the evolution of the voltage pulse. During the time scale of the energy relaxation processes that convey the detector back to its initial state, the detector is insensitive to additional photon absorption. This dead time limits the maximum count rates that can be achieved. On the other hand, some readout systems require a certain minimum pulse width due to their limitations of the input bandwidth. For time-correlated single-photon counting (TCSPC) applications, the time variance (jitter) between photon absorption and count event is an additional important property.

2.1.2 Preliminary work

SNSPD have been the topic of intensive research over the past decade [1]. Many efforts have been taken to improve the design towards an application-ready detector structure, mainly directed towards the increase in the practical detection efficiency of the nanowire devices. In the following, recent progress on this topic is collected with an emphasis on the most influencial approaches concerning the improvement of the spectral detection efficiency.

Optical coupling efficiency

The classical configuration is the coupling of the light normal to the substrate surface, either from the front or also sometimes from the back-side through the substrate. The first step towards an improved optical coupling was to fold the nanowire into a meander structure to cover larger detector areas of several μm^2. An SEM image of such a meander structure is shown in Fig. 2.2 a). The so-called filling factor is given by the relation between the linewidth w and the gap between the wire g and determines how much of the detector area is covered by the nanowire. This filling factor was increased up to 88% for a 90 nm wide wire [13]. However, a reduction of the critical current density with increase of the filling factor was observed, due to current crowding at the ends of the decreasing gaps. A further enhancement of optical coupling can be achieved by the integration of gold nano-antennae around the nanowire. With this approach, a device efficiency of 47% was shown at 1550 nm [14].

Optical fibers are often used to guide the light efficiently to the detector area, as shown in Fig. 2.2 b). Even so, focusing the light on the detector area is challenging. Approaches include nanopositioning of fiber end to detector at cryogenic temperatures by piezo stages with micro-optics [15] and lensed fibers, where a system detection efficiency of 24% at 1550 nm was demonstrated [16]. High detection efficiencies can be achieved also for coupling to multimode

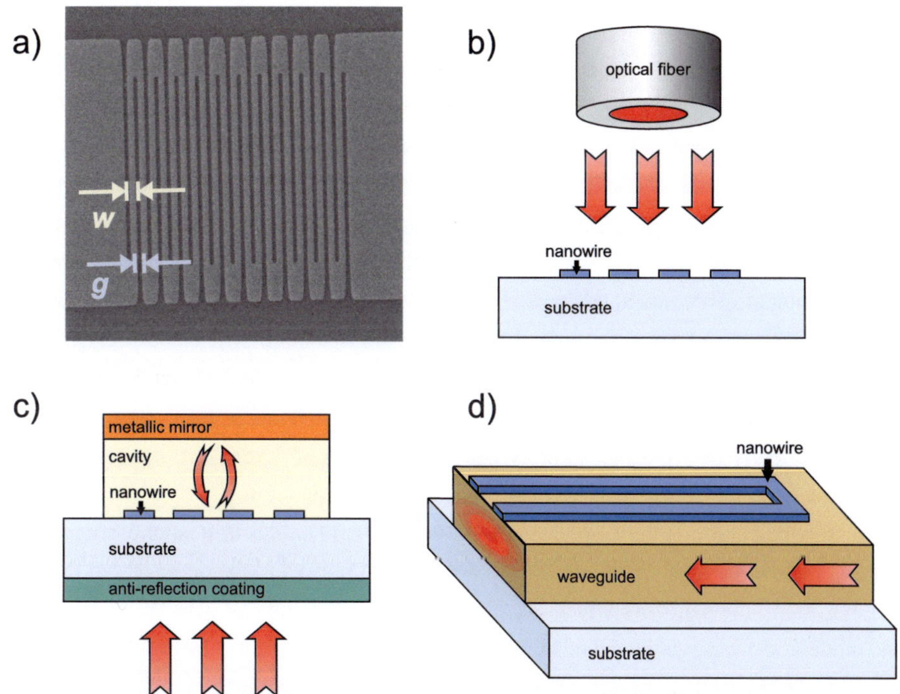

Figure 2.2: a) SEM image of a typical NbN SNSPD structured in meander shape. The light areas show the nanowire with linewidth w and the connections left and right. The dark areas show the substrate beneath the nanowire. b) Classical configuration of the coupling of the light normal to the substrate surface (light blue), where the light is guided to the detector (dark blue) by an optical fiber from the top. c) Configuration of an optical cavity (yellow) on top of a nanowire with back-side illumination. Once coupled into the cavity, the light is reflected back and forth between the substrate and a mirror, giving multiple chances to be absorbed in the nanowire. d) Configuration of a nanowire coupled to a waveguide (brown). In this case, the light propagates along the waveguide. The nanowire is patterned directly on top of the nanowire, parallel to the light, so no meander shape is neccessary.

optical fibers [17] which enables a broader wavelength range of the system. Other methods include gluing the optical fiber directly on the detector or even structuring an SNSPD on the face of the polished fiber end [18].

Absorption

If the light is successfully guided to the detector area, the detection efficiency is still limited by the low absorption probability of the thin superconductor. Since the device geometries are sub-wavelength, the total efficiency is not so easily discernible into filling factor times absorptance of a thin film. As a first approximation the thin film absorptance is nevertheless often used, which is for a typical NbN thin film roughly 30% [19]. To improve this ratio, the detector can be placed onto a substrate with back-reflection or a refractive index layer system which provides constructive interference at the desired operation wavelength. On a SiO_2/Si bilayer, a detection efficiency of 23.2% at 1310 nm with 1 kHz dark count rate was reported [20]. An integration into GaAs/AlAs Bragg mirrors was put forward with a detection efficiency of 18.3% at 1300 nm and 4.2 K [21]. The approach was taken one step further by embedding the nanowire into an optical cavity that allows for multiple chances of photon absorption (see Fig. 2.2 c)). Detection efficiencies of 57% at 1550 nm and 67% at 1064 nm at 1.8 K have been demonstrated on a 3 μm × 3 μm nanowire meander in an optical cavity with antireflection coating [22, 23]. Recently, for SNSPD made from WSi embedded in an optical cavity, system detection efficiencies of up to 98% at telecommunication wavelengths were reported, and a considerable 2.5% at 5 μm wavelength [24]. However, the operation temperature for this system at 120 mK limits the applicability.

Another recently emerging technology is the integration of SNSPD on optical waveguides. This is especially interesting due to its high integrability in quantum photonic circuits which are used for fundamental quantum experiments as well as application in quantum computing and quantum communications. Here, the light is not incident normal to the detector, but through the waveguide along the nanowire which is patterned on top (see Fig. 2.2 d)). The evanescent field of the mode in the waveguide extends into the NbN and causes a finite absorption probability. In this way, the approach is not limited by the low absorption rate caused by the thin film thickness, but by the interaction length. For NbN nanowire detectors on a GaAs waveguide, high detection efficiency (about 20%) at telecom wavelengths, high timing accuracy (about 60 ps), and response times in the ns range have been demonstrated [25]. By improvement of the modal matching in so-called travelling wave SNSPD, the detection efficiency can even be increased up to 91% at telecom wavelengths [26].

Intrinsic detection efficiency

The last factor, the intrinsic detection efficiency, is the property of main interest in this study. Many investigations have been directed towards the identification of contributions to the *IDE*, although the results seemed sometimes controversial.

One approach that has been taken is the reduction of the cross-section of the nanowire, i.e. the film thickness and the wire width. It has been shown experimentally for SNSPD made from NbN thin films on sapphire, that a decrease of the film thickness results not only in a broader spectrum of the detection efficiency, but also in an increase of the intrinsic detection efficiency of the nanowire [27]. Figure 2.3 shows the spectral dependence of the detection efficiency for NbN films of four different thicknesses from this reference to illustrate the dependence on film thickness. The presented values of film thickness are already close to the technological limit of the sputter deposition technique and can not be expected to decrease much further. A reduction of the nanowire width increases the spectral bandwidth in a similar manner, as has been shown

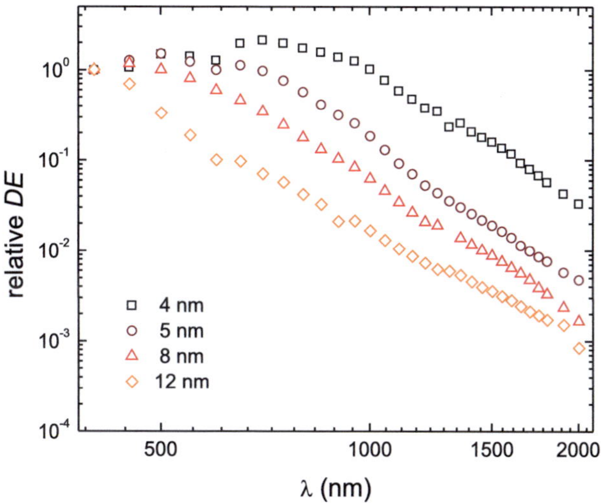

Figure 2.3: Spectral detection efficiency of several SNSPDs made from NbN films of different thicknesses [28]. The width of the wire as well as the measurement conditions were kept constant. The *DE* is shifted vertically so that all curves coincide at 400 nm as to better show the relative behavior. The shift of the cut-off wavelength towards the infrared range by reduction of the film thickness is clearly visible. The slight upturn in the low wavelength region is attributed to the wavelength-dependence of the absorptance. The measurements were taken at the DLR in Berlin.

for NbN SNSPD with widths below 50 nm [29]. For even smaller width of the nanowire, the signal amplitude of the devices due to the accompanying reduction of the critical current becomes increasingly demanding for the readout. The problem was partly solved by connecting nanowires of the detectors in parallel, where the switching in one strip due to a count event will induce an avalanche-like switching of the remaining strips [30, 31, 32]. In this way, devices with linewidths down to 30 nm can be realized [33]. Such devices have been termed superconducting nanowire avalanche photodetectors (SNAP) and provide short reset times due to the low inductance, but they are limited by strong afterpulsing [34].

Another approach was to investigate the influence of the superconducting film on the *IDE*. Many materials have been considered for the fabrication of SNSPD, where prominent ones are NbN and more recently NbTiN [35]. They offer relatively high critical temperatures so the devices can be cooled with liquid helium, are mechanically stable and can be deposited as thin films. To improve the spectral bandwidth, materials with smaller energy gap Δ but similar disordered nature were investigated as basis for SNSPD. TaN as a second nitride was proposed for its lower T_C value than NbN [36] and an increase of the cut-off wavelength by a factor of 1.3 was experimentally demonstrated [37]. Materials with even lower band gaps that were considered are amorphous NbSi, which demonstrated higher DE for wavelengths above 1550 nm relative to a NbTiN SNSPD [38], and amorphous tungsten silicide, where detection efficiencies of 19%-40% were achieved in the wavelength region from 1280 nm to 1650 nm [39]. Their critical temperature values below 2 K require that such devices have to be operated in the sub-Kelvin range, limiting their applications options. On the other hand, for detectors made from magnesium diboride, which has a relatively large band gap with T_C above 18 K, it was shown that infrared photons above 1560 nm can no longer be detected [40].

However, the band gap is not the only material parameter to influence on the intrinsic detection efficiency. For example detectors made from a pure Nb thin films, which have lower critical temperature values than NbN films, the detection efficiency was shown to be reduced and the spectrum was much narrower [41, 42]. This was attributed to the higher electron diffusion coefficient in Nb which hinders a local aggregation of quasiparticles.

Recently, it was suggested [43], that the geometrical layout of the SNSPD in the meanwhile 'classical' meander structure may lead to a reduction of the critical current of the device. A respective connection between filling factor and device performance has been put forward. Although first experimental verification of the proposed effect was published [44, 45], the extend to which it is applicable to actual detector structures is still under discussion.

2.2 Hot-spot mechanism and vortex-assisted photon detection

2.2.1 The hot-spot model

The current-assisted detection mechanism of a nanowire single-photon detector was originally proposed by Semenov et al. [46]. The model considers a photon with energy $E_{ph} = h\nu = hc/\lambda$ that is absorbed in a superconducting film. The energy of an optical or infrared photon is much higher than typical superconductor band gaps and immediatly causes the creation of quasiparticles upon absorption. The energy is distributed very fast from the initially absorbing electron by scattering with other electrons or phonons in an avalanche-like process. If the number of generated non-equilibrium quasiparticles is sufficiently high, the superconductivity will be locally destroyed and a normal-conducting zone appears around the point of incidence. The dynamics of this hot-spot predict a maximum size which depends on the energy of the photon, the quasiparticle multiplication efficiency and the diffusion channels. If this spot grows large enough to fill the whole cross-section of the wire, the device switches into resistive state. However, even if the spot is smaller than the cross-section and thus short-cut by a superconducting path, the redistribution of the current from the normal-conducting spot into these paths increases the supercurrent density. Once it surpasses the critical density, the whole cross-section switches temporally into the normal state, giving rise to a voltage appearing across the zone.

The dimensions of the cross-section of the wire and the ratio of bias current to critical current define a minimum spot size and thus a minimum energy E_0 of the photons that can be detected by this mechanism. For photons with lower energies $E_{ph} < E_0$, however, there is still a finite probability for detection due to multi-photon processes or fluctuation-assisted detections. This will be further discussed in section 2.2.2.

Although the principle of operation described in this model was successfully demonstrated, the predicted hot-spot sizes were too small to explain experimental evidence. The model was subsequently advanced to explain resistive state formation in a superconducting strip even though no normal spot appears [47]. The following section closely follows the description of the refined hot-spot model given in [48]:

Since SNSPD devices are made from very thin films where the thickness d is even below the coherence length ξ of the material, it is sufficient to consider a two-dimensional strip with a width w. The effective penetration depth $\lambda_{\text{eff}} = 2\lambda^2/d$ is very large and exceeds typical line widths at all temperatures, so that the influence of magnetic fields can be neglected and the distribution of the superconducting current is homogeneous (typical values for NbN thin films are $\xi \approx 4$ nm and $\lambda \approx 300$ nm). However, since $w > \xi$, it is possible for external vortices to enter into the current path which can cause fluctuation-induced dark count events. Those events are not included in the refined hot-spot model.

The average supercurrent density within the strip can be expressed as $j_S = 2en_Sv_S$ with v_S the mean Cooper pair velocity and n_S the mean pair density. A variation of the local Cooper pair density can occur only over distances larger than ξ along the current path. If a photon is absorbed in the film, the redistributed energy is breaking Cooper pairs, reducing their local density by δn_S. due to charge-flow conservations, the mean pair velocity has to increase locally to

$$v_S' = \frac{n_S}{n_S - \delta n_S} v_S. \tag{2.1}$$

The velocity follows the change in density on time scales almost instantaneous compared to energy relaxation times. Once the velocity exceeds the de-pairing value that corresponds to the critical current density $j_C = 2en_Sv_C$ without any excitations, the critical state is reached in the whole local area and superconductivity breaks down. The minimum change in the density of Cooper pairs respective to the mean pair density that produces a counting event is thus given by

$$\frac{\delta n_S}{n_S} = 1 - \frac{I}{I_c}. \tag{2.2}$$

Well below the transition temperature, the density of Cooper pairs can be approximated by the electronic density of states N_0 and the energy gap Δ: $n_S \approx N_0\Delta/2$. The total number of required generated quasiparticles $\delta N_{e,c}$ (twice that of broken Cooper pairs) inside the smallest necessary volume $V = \xi wd$ thus amounts to

$$\delta N_{e,c} = N_0 \Delta \xi wd(1 - I/I_c). \tag{2.3}$$

At a distance r from a photon absorption site, the concentration of non-equilibrium electrons $n_e(r,t)$ evolves in time due to the electron avalanche, described by the time-dependent multiplication factor $M(t)$, and the out-diffusion of the quasiparticles

$$n_e(r,t) = \frac{M(t)}{4\pi Ddt} \exp - \frac{r^2}{4Dt}. \tag{2.4}$$

After a certain time $t = \tau_{th}$, all non-equilibrium electrons have thermalized to the energy level Δ, thus becoming quasiparticles. At that time, already some of the initial energy is lost due to out-diffusion and recombination. The maximum number of quasiparticles is given by $M(\tau_{th}) = \varsigma E_{ph}/\Delta$, which defines the efficiency of the quasiparticle multiplication process, ς.

Integrating the number of quasiparticles over the volume V at the time τ_{th} when the number is at it's maximum, one arrives at

$$\delta N_e = M(t) \frac{\xi}{\sqrt{\pi D \tau_{th}}}. \tag{2.5}$$

Once this number surpasses $\delta N_{e,c}$ given above, the superconductor will switch into the normal state. Contrary to the original hot-spot model, the hot-spot is not completely normalconducting, but a zone with a reduced density of Cooper pairs. The nanowire switches to the normal state due to a combined effect of this local reduction and the current passing through the nanowire. Taking all together, the minimum photon energy necessary to create a count event is given by

$$E_{ph} = \frac{\sqrt{\pi}}{\varsigma} N_0 \Delta^2 wd \sqrt{D\tau_{th}}(1 - \frac{I}{I_c}). \tag{2.6}$$

For optical photons, the amplitude and duration of the signal output are defined by the evolution of the normal domain. They depend on bias current, the thermal transport conditions and the embedding circuitry and can be evaluated with the help of an electro-thermal feedback model [49, 50, 51]. Depending on those parameters, after reaching its maximum size, the domain can either shrink again and disappear or stay in the normal state until the bias current is reduced below a certain threshold (so-called latching).

2.2.2 Photon detection beyond the cut-off wavelength

From the hot-spot model described above, one would expect that only photons with sufficient energy can be detected. For photon energies above a certain minimum energy $E_{ph} > E_0$, but smaller than the plasma frequency, the intrinsic detection efficiency should equal the probability

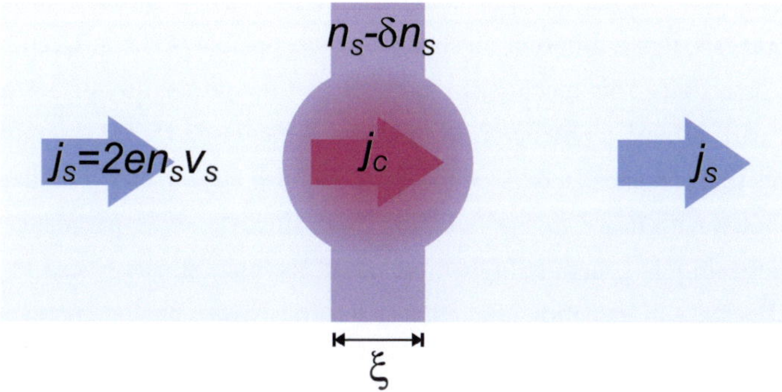

Figure 2.4: Schematic illustration of the formation of a hot-spot in a superconducting nanowire (blue) with a supercurrent $j_S = 2en_Sv_S$. Around the photon absorption site, the Cooper pair density n_S is reduced (red) by a fraction δn_S.

of photon absorption and not be dependent on the photon energy. For photons with $E_{ph} < E_0$, the detection efficiency should rapidly drop to zero. Indeed, the experimentally observed detection efficiency spectra typically show a so-called cut-off above a certain wavelength. In contrast to the expectation, for photon wavelengths exceeding this limit the detection efficiency is gradually reduced over a wide spectral range.

Many models were proposed to describe infrared photon detection within the hot-spot model. However, in the case of NbN SNSPD, they were shown to be not sufficient to explain the experimental evidence [28]: Constrictions of the nanowire cross-section [52] that are present due to the limitations of the fabrication processes are statistically distributed along the wire and lead to the possibility to locally detect photons with lower energy than the threshold. To explain the broad spectral range in which the experimental roll-off is observed, such constrictions would have to be by far large enough to be observed by scanning electron microscopy, which is not the case. Fluctuations in the superconducting order parameter [53], would have to be above 75% to describe the experimental data. Measurements by scanning tunneling microscopy found that such variations exist, but only in the range of 15% [54]. Both constrictions and fluctuations of the order parameter would also considerably limit the critical current of the devices to a much stronger extend what is generally observed. Furthermore, typical nanowire widths and hot-spot sizes are larger than possible grains of the superconductor, so no influence on the local detection efficiency should be expected from any non-uniform nature of the film.

Nevertheless, there is an observed reduction of the critical current in detector devices below the de-pairing critical current value (discussed in detail in chapter 5). The deviation appears at conditions, for which the entry of magnetic vortices into the nanowire is possible. Additionally, it was shown that the dependence of the intrinsic detection efficiency on film thickness is qualitatively different for photons below and beyond the cut-off energy [27]. While it is almost independent on film thickness for optical wavelengths, the intrinsic detection efficiency decreases quasi-exponentially with increasing film thickness for the case of infrared wavelengths. This leads to the hypothesis, that a different mechanism is responsible in the infrared range: the fluctuation assisted detection. This mechanism assumes that the probability of normal fluctuations in the nanowire parts around a photon absorption site is increased [55]. Generally speaking, all fluctuation events that can cause a dark count event can also appear with a higher probability at a site where a photon with $E_{ph} < E_0$ has been absorbed, since the order-parameter is then locally reduced. The probability is generally higher, the closer the photon energy was to E_0. This leads to a continuous decrease of detection efficiency as we go to lower photon energies.

Vortex nucleation and movement

An extensive study of various current-induced fluctuation phenomena in superconducting nanoscaled meander structures has been conducted by Bartolf et al. [56] to decide on the mechanism responsible for infrared photon detection. They come to the conclusion that thermally activated or quantum-mechanical-phase slips as well as the quantum-mechanical tunneling of vortices can be excluded as the dominant contribution. It should be however noted, that recently phase-slip centers were once again considered as switching mechanism for current values very close to the de-pairing critical current limit in the case of nanowires with bends [57]. The remaining two mechanism under consideration are the unbinding of vortex-antivortex pairs (VAP) [58] and the thermal excitation of single vortices across the edge barrier and the consecutive dissipative movement of the vortices across the strip. The potential barrier for the unbinding of VAP is much higher than that of a single vortex crossing [59], so that the latter is currently assumed to be the main source of dark counts and fluctuation assisted photon counts. The qualitatively different dependence of the intrinsic detection efficiency on the nanowire thickness described above can be well explained within the model of assisted photon detection by thermal excitation of magnetic vortices, which will be briefly discussed in the following:

If the lower critical magnetic field B_{C1} is exceeded by the magnetic self-field of the bias current, the intrusion of vortices into the nanowire is prohibited only by a potential energy barrier that is similar to the Bean-Livingston surface barrier [60]. This barrier is described as

$$U = \varepsilon_0 \left[\ln\left(\frac{w}{\pi\alpha\xi\sqrt{1+\left(\frac{I_B}{I_0}\right)^2}} \right) - \frac{I_B}{I_0}\left(\arctan\left(\frac{I_0}{I_B}\right) - \frac{\pi\alpha\xi}{w} \right) \right], \tag{2.7}$$

where the parameters $\varepsilon_0 = \Phi_0^2/(2\pi\mu_0\Lambda)$ and $I_0 = \Phi_0/(2\mu_0\Lambda)$ define the energy scale and the current scale, respectively. Φ_0 is the magnetic flux quantum, μ_0 the magnetic permeability and $\Lambda = 2\lambda^2/d$ the Pearl length [61], where in turn λ is the magnetic penetration depth of the superconductor and d,w the thickness and width of the nanowire. The parameter $\alpha = 4/\pi$ represents the smallest distance between the nanowire edge and the center of a nucleated vortex. The vortex

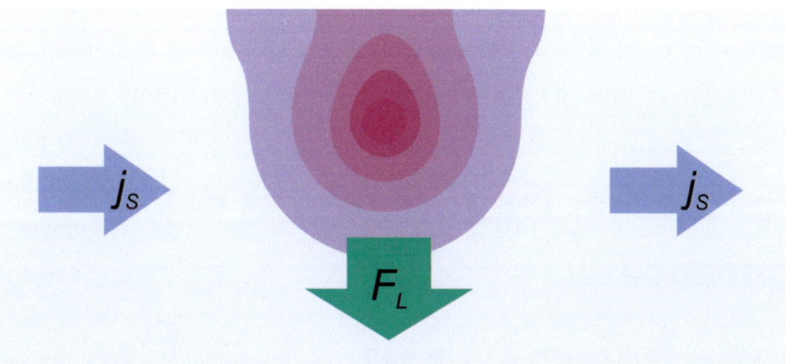

Figure 2.5: Illustration of the nucleation of a vortex at the edge of a superconducting nanowire. The color schematically shows the reduction of the Cooper pair density n_S, which reaches zero in the center of the vortex. The applied bias current exerts a Lorentz force F_L on the vortex directed to the opposite side of the wire.

can overcome this barrier either by thermal excitation or by tunnelling [62], but the probability for the latter becomes larger than thermal activation events only at temperatures below 1 K. Once the vortex is introduced into the nanowire, the Lorentz force will act on it due to the bias current and it will start to move across the wire. Since the energy dissipated in this case is similar to the energy of an optical photon, this will lead to a count event. Depending on the thermal activation energy, this can happen even without a photon absorption and is then considered to be a dark count event. If an infrared photon is absorbed, the order parameter is locally suppressed and consequently the potential barrier will be reduced by δU. The probability P for a vortex to be excited over the barrier is proportional to $I_B \exp(-U/k_B T)$ [62]. Defining Ω as the attempt rate with which the vortices try to overcome the barrier, the change in the probability δP due to the change in the potential energy δU is given by

$$\delta P = \Omega I_B \exp\left(-\frac{U}{k_B T}\right)\left[\exp\left(\frac{\delta U}{k_B T}\right) - 1\right]. \tag{2.8}$$

The intrinsic detection efficiency scales with the change in excitation probability, leading to a quasi-exponential decrease as the photon energy is reduced. This prediction fits well to the experimentally observed spectral dependence of the intrinsic detection efficiency [28].

In summary, the absorption of a photon leads to the formation of a hot-spot in a superconducting nanowire. In the refined hot-spot model, in this region the number of Cooper pairs is reduced and the remaining pairs have to accelerate to carry the supercurrent. For a successful detection event, this hot-spot has to surpass a minimum size for a certain period of time, before relaxation to the initial state. A good material for SNSPD application thus has to provide fast energy ralaxation rates to generate a large number of quasiparticles in a short time, but a low diffusion coefficient to retain high locality of the quasiparticle accumulation. To enable detection events even for low photon energies, it is generally favourable to have a low band gap, so that more quasiparticles are created from the same energy, and a low density of electronic states, so that the relative fraction of broken Cooper pairs is as high as possible.

2.3 Approaches for improvement of detector performance

Previous efforts described in section 2.1.2 have shown, that SNSPD-based detector systems can achieve very competitive detection efficiencies for radiation in the optical or near-infrared range. Solutions to overcome the limited absorption probability and the difficulties of the optical coupling to μm-sized devices are employed to great success. The main remaining open challenge for the development of SNSPD is the extension of the spectral bandwidth further into the mid- and far-infrared region. Progress in this direction is limited by the continuous decay of the intrinsic detection efficiency as the wavelength is increased. Therefore, the central aim of this work is to study factors influencing on the intinsic detection efficiency and dark count rate generation in SNSPD on a systematic basis, and to improve the spectral bandwidth towards the infrared range.

The minimum photon energy, which defines the point of the spectral cut-off in the hot-spot model, is given by eq. 2.6. The considerations discussed so far lead to the following approaches for the improvement of detector performance:

First, the influence of the properties of the superconducting material on the intrinsic detection efficiency is studied. In previous work, materials with small energy gap Δ and low density of electronic states N_0 have generally shown improved spectral bandwidths. A quantitative investigation including all parameter influences, however, does not exist so far. An additional problem arises from the fact that those parameters also depend on film thickness and nanowire width, which further hampers direct comparison between various published results. The task of this study is to investigate the intrinsic detection efficiency of identical SNSPD structured from a series of films with a gradual variation of material parameters. The gradual parameter change can be realized in the two-component material NbN by variation of the film stoichiometry. The influence of the NbN chemical composition on all relevant material properties is investigated by various experimental methods in chapter 3. Experimental results on the intrinsic detection efficiency spectra of SNSPD structured from those films will be discussed in section 6.2.

Energy relaxation and transport plays a significant role in the time-evolution of the hot-spot. The avalanche-like relaxation processes following the absorption of a photon in a superconducting film can be described in the framework of the Two-Temperature model. These processes define the intrinsic limit of the nanowire relaxation time, until it switches back into the superconducting state and can detect the next photon. The involved time constants are investigated with a frequency-domain technique in chapter 4. NbN film thickness and material composition and their influence on the energy relaxation are discussed to identify their limitations on the maximum SNSPD count rates.

The current dependent term in eq. 2.6 is especially interesting, because it suggests that by biasing the device closer to I_C, the minimum required photon energy could be reduced arbitrarily. We recall that I_C was defined as the current at which the velocity of the Cooper pairs is so high, that their the kinetic energy exceeds the pairing energy 2Δ. However, in real experimental situations it is observed that the device switches to the normalconducting state before I_C is reached. Typical values are around or even below $I/I_C = 0.5$. This limitation has recently attracted attention in SNSPD research, because an improvement of the ratio I/I_C would directly lead to a broader spectrum of the intrinsic detection efficiency. Possible mechanisms for this premature switching and possibilities to reduce their influence are explored on nanowires with simple geometries in chapter 5. Based on the results obtained in this investigation, an improved SNSPD layout is developed and results of the spectral detection efficiency measurements are presented and discussed in section 6.3.

The insights gained by the investigations described above are used to design SNSPD devices with improved properties, which is presented in the final chapter. Series of detectors with systematic variation of parameters are fabricated and characterized. Their dark count rates and detection efficiency in a broad spectral range are measured directly by a self-developed fiber-based setup and compared to reference samples that were made and examined under identical conditions. The results are then discussed in the framework of the hot-spot model.

3 NbN thin film technology and characterization

Niobium nitride is one of the most popular materials for SNSPDs. It has a relatively high critical temperature ($\approx 17\,$K in bulk material), which enables the operation of devices at $4.2\,$K, shows good mechanical and temporal stability and can be deposited into very thin films. Furthermore, the fast energy relaxation times and the low density of electronic states assure a significant reduction of the Cooper pair density upon absorption of a photon. At the same time the quasiparticle diffusion coefficient is relatively low, so that the formation of a hot-spot is possible.

It is also a suitable candidate to study the material properties of nanowire detectors due to the possibility to tune its superconducting parameters by variation of the deposition conditions. Superconducting and resistive properties of NbN found in literature almost always correspond to films where the deposition conditions were optimized to achieve maximum T_C. However, it will be shown that the dependence of these properties on thickness and width of nanowires as well as on slight variations of the deposition conditions makes it difficult to compare structures made from different films. Therefore, a full evaluation of material properties of the NbN thin films is necessary before patterning them into detectors.

The following section describes in detail the reactive magnetron sputtering process and the NbN thin film growth. The films are characterized by resistive measurements in the temperature range from $4.2\,$K to room temperature to extract the superconducting and normal state properties and their dependence on the chemical composition as well as film thickness are studied. It will be shown that the chemical composition can be tuned continuously by controlled variation of the deposition parameters.

Further properties that are relevant for the detector performance are extracted by complementary measurement techniques: Measurements in high external magnetic field allow conclusions about the disorder level of the films and give access to the diffusion coefficient D and the coherence length ξ_{GL}. The energy gap and its temperature dependence is probed by THz transmission spectroscopy. Finally, the optical constants of the material are studied by ellipsometry to further investigate the film structure and conclude on the absorption efficiency of the films. The structuring of nanowire detectors from the films and the measurement of detector performances in dependence on the material composition will be presented later on in chapter 6.

3.1 Influence of deposition current on NbN stoichiometry

The deposition of NbN films by reactive sputtering has been investigated extensively in the '80s-'90s, where the interest was to use the material as base electrode in tunnel junctions [63]. The deposition process is very sensitive and difficult to control due to the hysteretic effect that occurs with the addition of the reactive gas [64]. The stoichiometric composition of the compound Nb_xN_{1-x}, and in consequence all material properties, change with the variation of reactive magnetron sputter conditions like partial pressures, gas flow and discharge current [65, 66]. Highest critical temperatures are achieved for near 1:1 stoichiometric composition [67].

3.1.1 DC reactive magnetron sputtering

The basis of the thin film deposition is the sputtering of a Nb target in a vacuum chamber, assisted by a cooled magnetron to increase the sputter rates. A typical current-voltage characteristic of a pure Argon plasma is shown as open circles in Fig. 3.1. Below a certain threshold current, the plasma is unstable and the voltage increases erratically, with high temporal instability. This region is unsuitable for deposition. For higher currents, above approximately 80 mA, the plasma displays an almost linear, ohmic dependence, where the slope is defined by the flow and the pressure of the argon gas. The sputter rate of the Nb increases with the sputter current. These are the conditions used for the deposition of pure Nb thin films. The chamber is preconditioned by sputtering Nb for several minutes. During this time, the current is swept slowly to ensure a homogeneous pre-cleaning of the target by removing the surface layers that have been exposed to ambient atmosphere.

For the deposition of NbN, nitrogen as additional reactive gas is introduced into the atmosphere. The nitrogen molecules are broken up in the plasma into highly reactive radicals. The current-voltage characteristic is very sensitive on the partial pressure and becomes non-monotonic and hysteretic (filled circles in Fig. 3.1). For low currents the voltage rises almost parallel to the pure Ar dependence, but with a voltage offset of 70 V. After reaching a maximum, it starts to decrease until both curves coincide. The dependence can be qualitatively explained by the competition of the chemical reaction of the nitrogen radicals with Nb and the sputtering of target surface layers. The exothermal reaction can only occur in the presence of a third body due to energy conservation and thus takes place only on the target surface. The reaction rate depends on the nitrogen partial pressure. The change in discharge current I_{SP} influences only on the rate of the surface layer sputtering. For low currents, the sputter process is slower than the chemical reaction and the deposited film has a high nitrogen content. The offset in voltage is due to the lower emission rate of secondary electrons from the target if it is covered by a nitride layer. For high currents on the other hand, Nb is sputtered from the surface faster than it can react with the nitrogen, causing a discharge characteristic very similar to the one for pure Nb sputtering. Su-

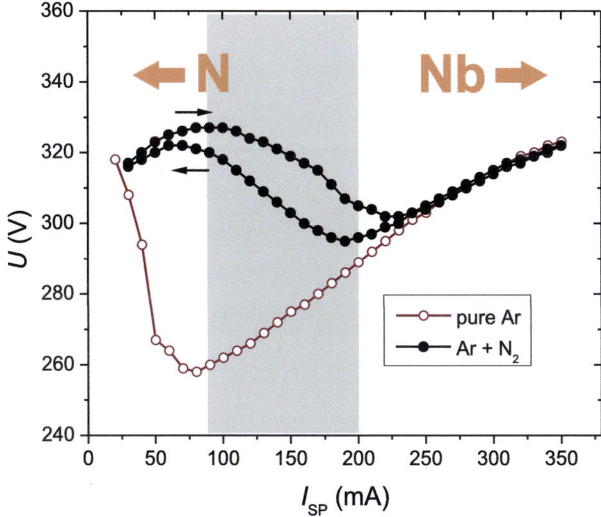

Figure 3.1: Discharge characteristic of the magnetron plasma for the deposition of NbN films. For a pure argon plasma (open symbols), the discharge voltage U rises continually with the sputter current I_{SP}. If nitrogen is added into the atmosphere (closed symbols) the dependence becomes non-monotonic and for the increasing and decreasing current branches (black arrows) hysteretic. Superconducting NbN films are deposited at conditions marked by the gray area, where the films have an excess amount of nitrogen for low I_{SP} and a nitrogen deficiency for high I_{SP}.

perconducting NbN is formed in the intermediate region, where the discharge voltage decreases with I_{SP} (gray area in Fig. 3.1). The position and steepness of this transitional area depends on the nitrogen gas flow: Increased N_2 flow shifts the transition to higher currents and it becomes steeper. For well-reproducible deposition conditions it is favorable to work at nitrogen flows for which the transitional range is as wide as possible, so that they can be stabilized at any particular deposition current.

The NbN films discussed in this work were deposited at conditions that correspond to the upper branch of the current-voltage characteristic inside the grey region shown in Fig. 3.1. The partial pressures were $p_{Ar} = 2.5 \cdot 10^{-3}$ mbar for Argon and $p_{N2} = 0.29 \cdot 10^{-3}$ mbar for nitrogen. The deposition rate was evaluated for several sputter currents by test films that were deposited on heated sapphire substrates. The test films were then patterned and the film height measured by profilometry. The obtained thickness values correspond well with film thicknesses obtained by ellipsometry and AFM-measurements (see section 3.5).

The evaluated deposition rates r_{dep} for the transition region (100 – 200 mA) are shown in Fig. 3.2. Generally, they should increase proportionally to the power: $r_{dep} \propto U I_{SP}$. However, since the discharge voltage U hardly changes in the range under consideration, the dependence

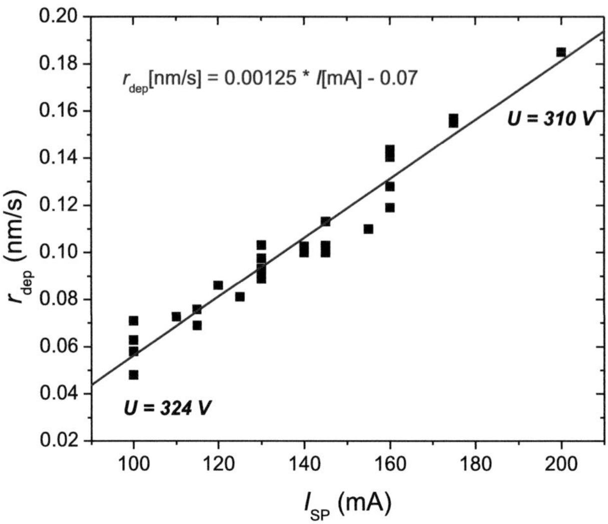

Figure 3.2: Deposition rates of NbN films evaluated at deposition conditions covering the gray area in Fig. 3.1. In the considered range the rate is to a good approximation proportional to I_{SP}. The respective linear fit is displayed as the brown solid line.

$r_{dep}(I_{SP})$ can be approximated as linear (solid line in Fig. 3.2). The rates vary between 0.06 nm/s and 0.18 nm/s.

3.1.2 NbN thin film growth

The growth of NbN, especially in the case of thin films, depends strongly on the choice of substrate. NbN with optimal T_C has a face centered cubic structure with a lattice constant $a_0 = 4.4$ Å [66]. Generally, NbN grows in a granular manner, where the grains themselves have nanocrystalline structure. The non-metallic grain boundaries act like a tunnel barrier and are the cause of the high resistivity. They can also lead to a negative dependence of the resistance on temperature [68]. Since the size of this grains and the properties of the grain boundaries depend on the growth conditions, a wide range of resistivity values can be obtained for NbN films. On the other hand, if a good lattice matching and good growth conditions are provided, the films can be almost single-crystalline [66]. Substrates that are commonly used include Si with native oxide layer [69], with a significantly larger lattice constant $a_0 = 5.4$ Å, and polished sapphire [70]. Epitaxial film growth has been demonstrated on the well-matched substrates MgO (100) with $a_0 = 4.2$ Å[66] and 3C-SiC (100)/Si(100) substrates, which are suitable for THz applications [71]. The growth conditions can be further improved by heating of the substrate. This enhances the surface mobility of the freshly sputtered particles and leads to fewer defects in the lattice structure. The critical temperature of films deposited at elevated temperatures can be several Kelvin higher than at room temperature.

The NbN films described in this work are deposited at polished R-plane sapphire substrates with a lattice constant of $a_0 = 4.8\,\text{Å}$. The material is a good compromise between reasonable lattice matching on the one side and mechanical and temporal stability on the other side. The substrates are cleaned carefully prior to deposition in n-hexane to dissolve organic materials and then with the aprotic solvent aceton and protic solvent isopropanol to remove any remaining material. For deposition the substrates are placed on a sample holder at room temperature. Once under vacuum, the holder is heated slowly up to a temperature of 750°C, leading to a substrate surface temperature of approximately 650°C.

The deposition of the first few atomic layers of the film are most strongly influenced by the lattice mismatch of the substrate and the film. As the film thickness increases, those conditions change during the deposition, leading to a gradual change of the stoichiometry and crystal structure between initial layer and upper layers. This means that a thin film, for which the deposition was stopped after the initial layer, is effectively made of a slightly different material than a thicker, longer deposited film. This change in material is reflected in the critical temperature T_C of films deposited at the same sputter conditions, but with different thickness (Fig. 3.3). Unlike pure Nb films, T_C rises with the film thickness without reaching the bulk value of $\approx 17\,\text{K}$, even for relatively large film thicknesses of 50 nm. For very thin film thicknesses (approximately for $d < 10\,\text{nm}$), however, there is an additional effect that strongly reduces the T_C and leads to the strong suppression shown in Fig. 3.3. This has been reported for Nb and NbN thin films [72, 73] and can be well explained as a proximity effect of the boundary layers on the NbN film [48]: The proximity effect was first proposed by Cooper [74] and describes the influence of normal-conducting layer on a superconductor in direct contact. The superconductor's energy gap is reduced in close vicinity of the interface area, i.e. on the length scale of the coherence length ξ_0. If the thickness of the superconductor is in the same order of magnitude or even lower than ξ_0, this leads to a reduction of the critical temperature of the film. This approach has been extended in [75] to describe bilayers in the diffusive limit with non-ideal interlayer interfaces. The case of a NbN thin film can be described in the framework of this model as a normalconductor-superconductor-normalconductor (NSN) layer system, where the critical temperature of the superconductor T_C^0 is reduced to T_C as

$$\ln\left(\frac{T_C^0}{T_C}\right) = \frac{\tau_N}{\tau_S + \tau_N}\left[\Psi\left(\frac{1}{2} + \frac{(\tau_S + \tau_N)\hbar}{k_B T_C \tau_S \tau_N}\right) - \Psi\left(\frac{1}{2}\right) - \ln\sqrt{1 + \left(\frac{\tau_S + \tau_N}{\tau_S \tau_N \omega_D}\right)^2}\right], \qquad (3.1)$$

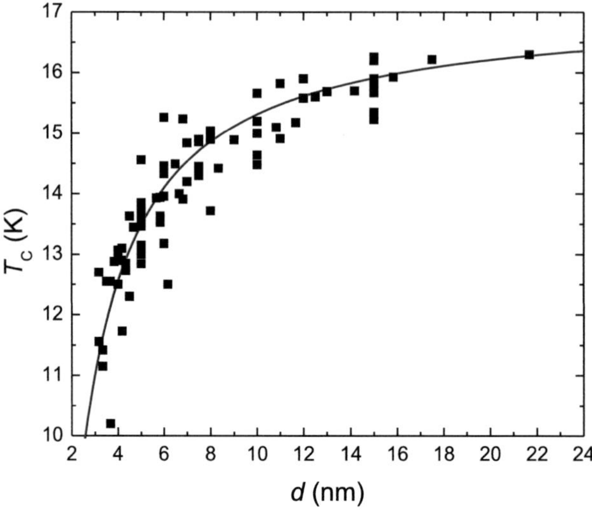

Figure 3.3: Dependence of critical temperature T_C of NbN films deposited at conditions for optimal T_C-values on the film thickness. The solid line is a fit of the proximity effect described by eq. 3.1 to the data.

where $\Psi(x)$ is the digamma function, ω_D is the Debye frequency of the superconductor and k_B and $\hbar$ are Boltzmann's and Planck's constants. The parameters τ_S and τ_N are defined as

$$\tau_S = \pi \frac{d_S}{v_S} \rho_{int} \quad , \quad \tau_N = 2\pi \frac{v_N d_N}{v_S^2} \rho_{int},\tag{3.2}$$

with $d_{S,N}$ the thickness and $v_{S,N}$ the Fermi velocity of the superconductor and the normal conductor, and ρ_{int} the resistivity of the interface between them.

The model predictions fit very well to the experimental $T_C(d)$ values (solid line in Fig. 3.3), confirming the assumption of the NSN layer system. The normalconducting layers can be identified as the interface layer between the film and the substrate, and the surface layer of the film, as schematically shown in Fig. 3.4. Once in contact with ambient atmosphere, the topmost layer of the NbN film oxidizes, leading to the formation of niobium oxide. The interface layer can also be partly oxidized, depending on the substrate used, and is additionally disordered due to the initial lattice mismatch. The layered composition and the nano-crystalline structure of NbN thin films have been confirmed by high resolution TEM measurements [19] and EELS analysis [69]. A similar proximity effect can also occur on structured samples with a small width w, where the edges of the structure act as the normalconducting zone due to damaging or oxidation. In this case, the NSN layer system is parallel to the substrate surface.

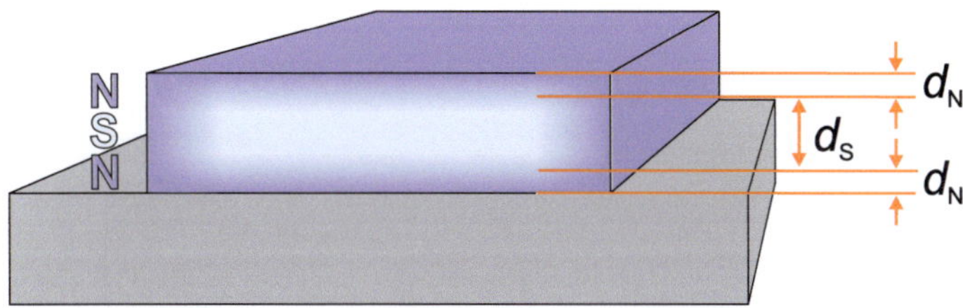

Figure 3.4: Schematic illustration of the proximity effect in a superconducting bridge. The material is superconducting in the core (light blue), but normalconducting at surface layers with thickness d_N (dark blue). Due to the proximity effect of the normal layers on the superconductor, the critical temperature of the whole bridge is reduced if d_N becomes significant with respect to the thickness. The remaining effective superconducting film thickness is d_S. The same effect is observed laterally, when the width of the superconducting bridge is very small.

Below film thicknesses of ≈ 2.5 nm, the reduction of T_C becomes stronger than even the proximity effect prediction. Likewise, the resistivity of the films increases with a stronger dependence and the material transits to an insulator. This is the range where the film does not yet cover the whole surface of the substrate. The NbN grows in islands that are randomly distributed on the substrate surface. For very low deposition times these islands are isolated from each other so that no conductive path can be formed. Once they grow enough to come into contact, a current can be pathed but the non-metal boundary between the islands cause the high resistivity.

Since the stoichiometric composition of the sputtered NbN films and the respective film parameters strongly depend on the thickness, it is of importance that films used for a comparison of material properties are deposited with the same film thickness.

3.2 Dependence of NbN film properties on deposition current

After the determination of the sputter rates, it is possible to make a series of films with almost constant thickness but each deposited at a different sputter current I_{SP}. The nominal film thickness is typically achieved with an accuracy of ± 0.5 nm. Within such a series it is then possible to observe how the sputter current influences on the the material composition by analysis of the film parameters. Due to the difference in film growth conditions as described above, the results depend on the actual film thickness. Therefore, several series of NbN films on sapphire substrates are made at miscellaneous film thicknesses in the range between $4 - 10$ nm.

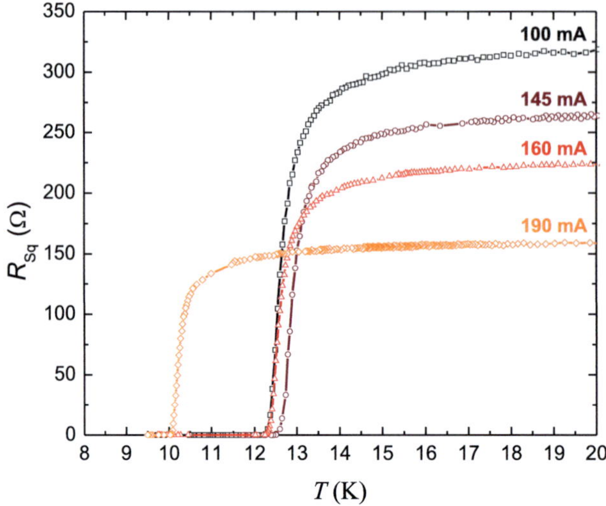

Figure 3.5: Square resistance R_{Sq} in dependence of the temperature in the region of the superconducting transition for 4 nm thick NbN films deposited at varying I_{SP}. The critical temperature of each film is defined where the resistance reaches zero within the measurement accuracy.

3.2.1 Critical temperature

The first characterization of the films is made on dummy-samples that were deposited simultaneously with the films that were patterned later on. This allowed the use of invasive contacting for the resistive measurements. The square resistance R_S of each film was measured over the full temperature range from room temperature to liquid helium at 4.2 K. The resistance varies usually only slightly with the temperature in the range 30 – 300 K but expresses a distinctive transition as the sample reaches the superconducting state. Fig. 3.5 shows several superconducting transitions from the series of 4 nm thick films deposited at different sputter currents I_{SP}. Coming from high temperatures, the drop in resistance starts slowly but becomes increasingly steeper until it finally reaches zero. Various different definitions of the critical temperature T_C can be found in the literature. The $R(T)$-dependence of NbN can be well described by a Cooper pair fluctuation model developed for two-dimensional superconductors [76]. In this model, the definition of T_C almost coincides with the highest temperature for which the resistance has vanished completely. We therefore evaluate T_C as the highest temperature for which the resistance of the film is below the measurement accuracy of the system.

For the films deposited at different I_{SP} in Fig. 3.5, a first indication of a change in the stoichiometric composition of the NbN is given by the apparent change in T_C. The dependence of T_C on the sputter current was evaluated in this way for three sample series of 4, 5 and 10 nm film thickness. The result is shown in Fig. 3.6. Generally, the T_C-values of the thinner films are below the values of the thicker films, reflecting the thickness-dependence described in the last section. Within each series, there is a maximum T_C value, which appears at the same I_{SP} for all series (for the deposition parameters presented here at 145 mA). This is the sputter current for which usually the NbN deposition is optimized. Therefore the material composition of films deposited at this current is termed 'standard composition' for the purpose of this discussion and marked by the blue bar. Going to either higher or lower I_{SP}, the critical temperature decreases. The relative behavior is similar for all series: For low I_{SP}, T_C is reduced gradually, up to almost 1 K below the maximum for the 80 mA film. For even lower sputter currents, the plasma becomes too unstable for deposition. For I_{SP} above the optimum, the decrease is much steeper. For currents above 200 mA, this strong dependence leads to a poor reproducibility and the T_C-values become too low for the fabrication of practical samples.

3.2.2 Resistivity

From the dependence of T_C, one could conclude that changing from the maximum to either lower or higher I_{SP} the material composition changes in the same direction. This can be excluded by looking at the resistivity of the samples. The bottom graph in Fig. 3.6 shows the resistivity values ρ_{30} measured on the same three film series as described above. The values were taken at 30 K, well above the superconducting transition but close enough to T_C that we expect no major difference between this value and the normal state resistivity at superconducting temperatures. It can be observed that for a thicker film, the resistivity values are always lower than for a thin film. This can be attributed to the better crystalline quality of the upper layers of the film that allow for the formation of more uniform NbN with larger electron mean free path. But for all film thicknesses, there is also a clear decrease in resistivity with increasing sputter current I_{SP}. Compared to the standard composition (marked by the blue bar), the resistivity is about 30% larger for the 100 mA films, and almost 20% lower for the 180 mA films, exact values depending on the film thickness. This supports the assumption that for increasing sputter currents, the Nb content of the film is higher. However, for sputter currents up to 200 mA, the resistivity values of the films are still much higher than what would be expected for a film made of pure Nb ($\approx 17\,\mu\Omega$cm for a 9 nm thick Nb film [77]).

The assumption is further supported by the dependence of the residual resistivity ratio (RRR) which is evaluated as the ratio between resistivity at room temperature to the normal state resistivity close to the superconducting transition (here defined at 30 K): $RRR = R_{300}/R_{30}$. Figure 3.7 shows the resistance of two NbN films in the whole temperature region from 4.2 K to room temperature. For a film with metallic behavior, the temperature coefficient is positive ($RRR > 1$) like for the 175 mA curve. Similar to the resistivity, the RRR-value of a NbN thin film is also dependent on the thickness and decreases for thinner films. For NbN films with standard composition below 10 nm thickness, there is a cross-over to a negative temperature coefficient ($RRR < 1$) like for the 100 mA curve. Figure 3.8 shows the dependence of the RRR-values on the sputter cur-

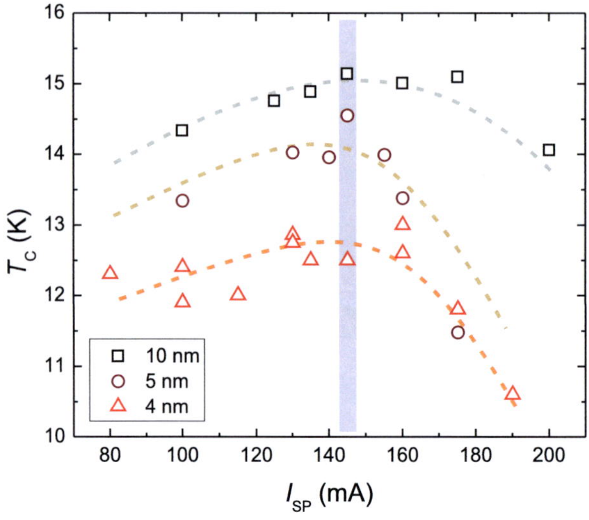

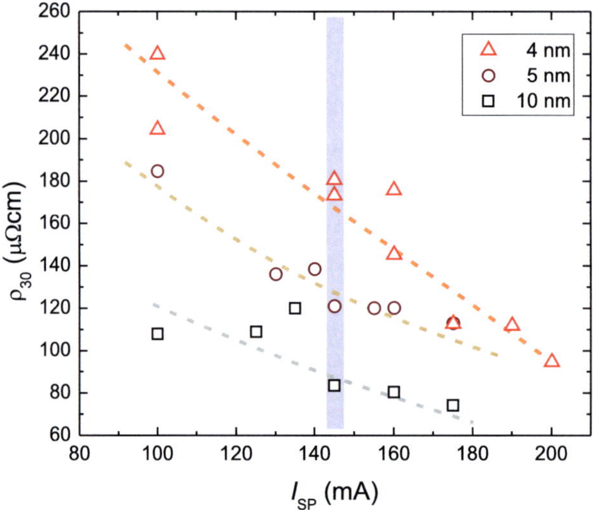

Figure 3.6: Top: Dependence of critical temperature T_C on sputter current I_{SP} at which the NbN film was deposited for three series of different film thickness. There is an optimum for all thicknesses at 145 mA (blue bar). As the thickness is reduced, the whole dependence shifts to lower T_C values. Bottom: Dependence of resistivity taken at 30 K, ρ_{30}, on sputter current I_{SP} for the same films. It decreases continually with I_{SP}. For increasing film thickness, the dependence becomes weaker and shifts to lower ρ_{30} values.

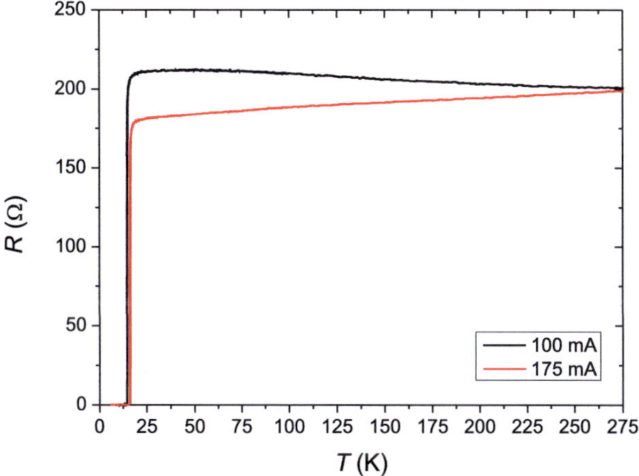

Figure 3.7: Resistance in the temperature range from 4.2 K to room temperature for two samples with 10 nm film thickness deposited at different I_{SP}. One sample shows a positive and the other a negative temperature coefficient, representing the cross-over from $RRR < 1$ to $RRR > 1$ in the series by varying I_{SP}.

rent for two sample thicknesses. Due to the thickness-dependence, the values of the 4 nm thick sample series are in all cases below those of the 10 nm thick series. In both cases, RRR increases continuously with I_{SP} in the direction of the metal-like behavior.

3.2.3 Critical current

For the determination of the critical current density, the films need to be patterned into bridges of defined width and length. In section 3.1.2 it was discussed that in patterned samples, the proximity effect can lead to a reduction of the superconducting properties if the width of the bridges is very small. To reduce this effect for the purpose of material characterization, the bridges are made several µm wide. Smaller bridges with widths in the sub-µm regime will be discussed in detail in chapter 5. The films are patterned by a standard one-step photolithography process with subsequent reactive ion etching.

Typical IV-curves from the series with 10 nm film thickness measured at 4.2 K are shown in Fig. 3.9. At this temperature, all curves are hysteretic and show a sharp jump of the microbridges from superconducting to normal state. The current at which this jump is observed defines the critical current I_C. The hysteresis current (the current at which the bridge returns into superconducting state with decrease in applied current) is determined by equilibrium between the Joule heating power dissipated in a microbridge and the efficiency of thermal flow through film-

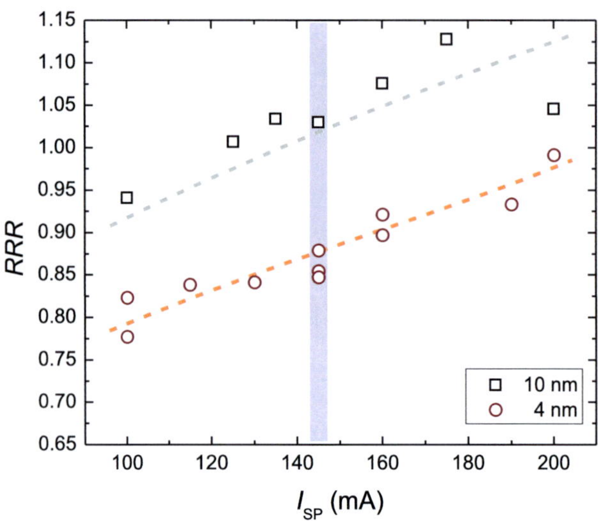

Figure 3.8: Dependence of the residual resistivity ratio RRR on the sputter current I_{SP} of the thin film deposition for two sample series of 10 and 4 nm thickness.

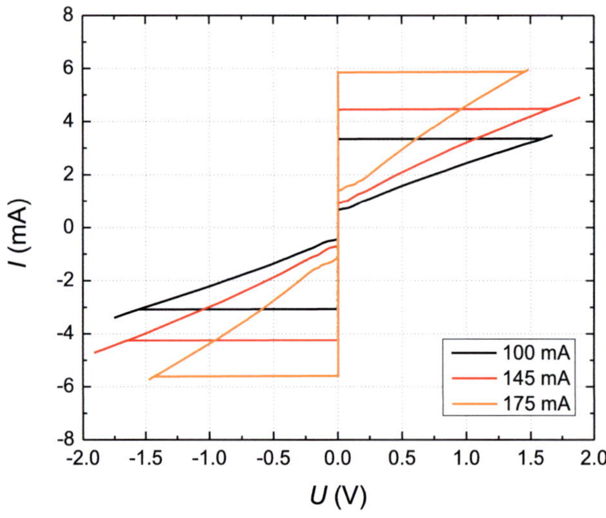

Figure 3.9: Current-voltage characteristics measured at 4.2 K for three selected samples from the series with 10 nm film thickness. The films were deposited at the sputter currents indicated in the legend.

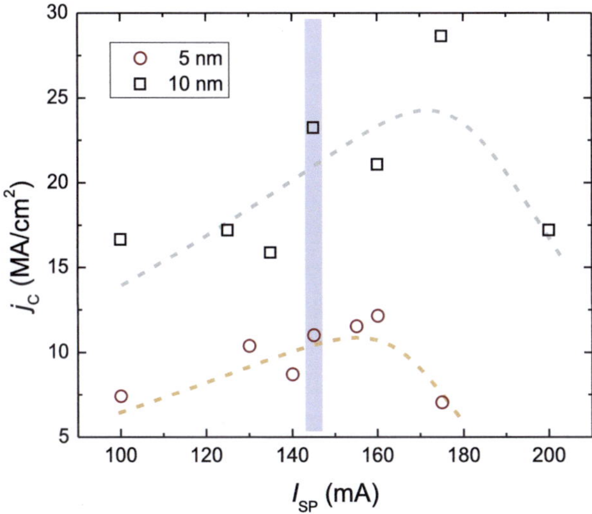

Figure 3.10: Critical current density measured at 4.2 K on bridges made from NbN films deposited at the respective I_{SP}. Two sample series of 5 and 10 nm film thickness are shown. The blue bar marks the I_{SP} for the NbN standard composition.

substrate interface [78]. The critical current density j_C is evaluated with the bridge width w and the film thickness d as $j_C = \frac{I_C}{wd}$ and averaged over all bridges structured on the same film.

The dependence of j_C on the sputter current of the film deposition is shown in Fig. 3.10 for two film thicknesses. Similar to the dependence of T_C, there is a moderate rise for low I_{SP}, a maximum value, followed by a relatively strong decrease for large I_{SP}. The maximum value, however, is not at the same sputter current as for T_C (145 mA, blue bar), but at slightly higher $I_{SP} = 160 - 180$ mA. The dependence of j_C of nanowire detector structures made from NbN films of varying chemical composition and the comparison to the theoretical de-pairing critical current will be further discussed in section 6.2.

The evaluation of the resistive properties of the normal state as well as the superconducting parameters show, that their gradual change can be achieved by variation of the sputter current I_{SP}. The superconducting parameters T_C and j_C both show a maximum in the considered region, which is at 145 mA for T_C and at roughly 170 mA for j_C. The NbN thin films growth is granular and the properties are dominated by the inter-grain boundaries which can have a chemical composition that is different from the grains themselves. The reduction in ρ_{30} and increase in RRR with increasing I_{SP} indicates a higher Nb content in the grain boundaries with metal-like properties. All parameters change in the direction towards pure Nb values, but even for the highest considered I_{SP} differ significantly from those. On the other hand, for low I_{SP} the grain bound-

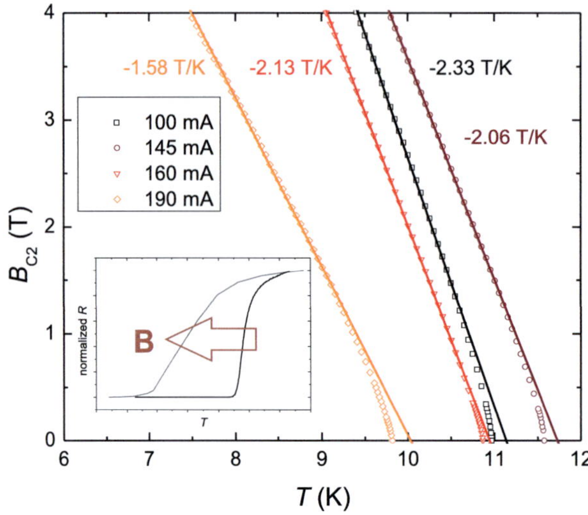

Figure 3.11: Temperature-dependence of second critical magnetic field B_{C2} for four 4 nm thick NbN films deposited at the sputter currents indicated in the legend. The solid lines are linear fits to the data with the exclusion of the points very close to T_C with the respective slope given next to each line. The inset shows the superconducting transition of a sample without magnetic field (black) in comparison to a high applied magnetic field (gray).

aries grow with enhanced nitrogen content and become increasingly isolating, resulting in high resistivity and low *RRR* values.

3.3 Second critical magnetic field

Additional material properties of the NbN thin films can be obtained prior to structuring by evaluation of the magnetic properties. The first and second critical magnetic fields H_{C1}, H_{C2} and their temperature-dependence allow conclusions on the structure of the material. The electron diffusion coefficient D, which enters the calculations of the hot-spot dynamics (see section 2.2.1), is obtained from $H_{C2}(T)$. Furthermore, the Ginzburg-Landau coherence length $\xi_{GL}(0)$ can be extracted from the extrapolation of $H_{C2}(T)$ to zero temperature. Since all these parameters change with the chemical composition of the films, they are dependent on the film thickness d as well as on the sputter current at film deposition I_{SP} and need to be evaluated separately. The measurements were conducted in a separate setup equipped with a superconducting magnet, where a field perpendicular to the sample surface up to 4 T can be realized. A temperature controller adjusts the sample temperature with an accuracy of about 5 mK in the range from 3.7 K to T_C. Details of the measurement setup can be found in [79].

The introduction of the magnetic field reduces the critical temperature. However, when looking at a granular material as NbN, the shape of the resistive transition itself changes (see inset of Fig. 3.11) [80]. The reduction of R_N begins at the same temperature, but the transition becomes much broader and reaches zero only at much lower temperatures. For the purpose of this measurements it is therefore sensible to define the critical temperature at the middle of the transition, i.e. at the temperature T for which $R(T) = R_N/2$. Correspondingly, we define the second critical magnetic field B_{C2} as the magnetic field, for which - at a fixed temperature - the resistance dropped to $R_N/2$.

Figure 3.11 shows the temperature-dependence of B_{C2} for several 4 nm thick NbN films deposited at sputter currents as indicated in the legend. The curves are shifted in x-direction to each other due to their difference in the critical temperature at zero field as described in the last section. The dependence is linear over almost the whole measurement range. However, close to T_C, the measurement points drop below the linearity. This behavior is commonly observed in NbN and has been attributed to the granular structure of the material [48]. The films grow in grains that are enclosed by a layer of NbN with excess nitrogen content with a lower T_C. The size of these grains depends strongly on the deposition conditions and the surface mobility of the sputtered particles on the substrate. For temperatures $T \ll T_C$, the Ginzburg-Landau coherence length ξ_{GL} is typically much smaller than such grains and magnetic vortices can form almost the same Abrikosov's lattice as they would in a uniform material. Close to T_C, however, ξ_{GL} diverges and eventually exceeds the grain size. This leads to a distortion of the vortex lattice and effectively to a reduction of the second critical magnetic field B_{C2}.

The derivative of the second critical magnetic field dB_{C2}/dT was thus taken from the linear part of the dependence. The electronic diffusion coefficient D is in the dirty limit ($l \ll \xi_0$, with ξ_0 the BCS coherence length) given by

$$D = -\frac{4k_B}{\pi e}\left(\frac{dB_{C2}}{dT}\right)^{-1}. \tag{3.3}$$

The calculated values are shown in the lower graph of Fig. 3.12 for several 4 nm thick films. The value for the standard film with 145 mA is 0.53 cm^2/s, similar to what is reported in the literature. With increasing sputter current, the diffusion coefficient rises, up to 0.69 cm^2/s for the 190 mA film. Note that this value changes also with the film thickness, since the material composition changes as described in the previous section. For standard composition the diffusion coefficient rises to 0.60 cm^2/s for a 6.5 nm thick film and up to 0.77 cm^2/s for a 10 nm thick film.

To acquire the value of $B_{C2}(0)$, the full temperature-dependence needs to be known. Although the linear B_{C2}-dependence holds true for a large temperature-region below T_C, the dependence should eventually saturate for very low temperatures. The dependence has been described for a

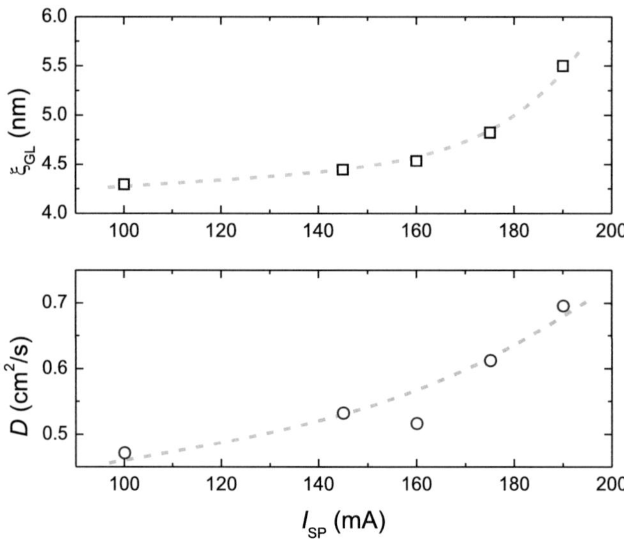

Figure 3.12: Coherence length ξ_{GL} (upper graph) and electron diffusion coefficient D (lower graph) in dependence of the sputter current I_{SP} for a series of 4 nm thick NbN films.

superconductor in the dirty limit in [81] which agrees very well with high magnetic field measurements. As a good approximation, $B_{C2}(0)$ can be extrapolated as

$$B_{C2}(0) = 0.69 \frac{\mathrm{d}B_{C2}}{\mathrm{d}T} T_C. \tag{3.4}$$

With this, the Ginzburg-Landau coherence length $\xi_{GL}(0)$ can be calculated according to

$$\xi_{GL}(0) = \sqrt{\frac{\Phi_0}{2\pi B_{C2}(0)}}, \tag{3.5}$$

where $\Phi_0 = h/2e$ is the flux quantum. The values are shown in the upper part of Fig. 3.12 for the same films as for the diffusion coefficients and display a quite similar behavior: With increasing sputter current, the coherence length becomes larger, rising from 3.56 nm up to 4.57 nm. In the case of a dirty superconductor, $\xi_{GL}(0) \propto \sqrt{\xi_0 l}$, where ξ_0 itself is inversely proportional to T_C. So the film with highest T_C at 145 mA should display the minimum ξ_0 value and it should rise to either higher or lower currents. The fact that $\xi_{GL}(0)$ rises monotonically suggests that the electron mean free path l also increases with the sputter current. This assumption is further supported by the increasing electron diffusion coefficient D.

In conclusion, the measurement of the temperature-dependent critical magnetic field of unstructured NbN films has confirmed the granular structure of the material. The electron diffusion coefficient D and the Ginzburg-Landau coherence length $\xi_{GL}(0)$ have been extracted for a series of 4 nm thick films with varying sputter current. Both parameters increase with increasing Nb content, moving towards - but not reaching - the values of pure Niobium.

3.4 Superconducting energy gap

The previous chapters have shown that the superconducting properties of the NbN thin films change with the variation of the sputter current. This chapter focuses on the influence of I_{SP} on the superconducting energy gap Δ. In the BCS theory, the gap at zero temperature is directly connected to the critical temperature T_C via the relation

$$2\Delta(0) = 3.528 k_B T_C, \tag{3.6}$$

where k_B is the Boltzmann constant. For typical T_C values of the NbN films considered here, energy gap values calculated in this way are sevaral meV. However, for many superconducting materials this relation does not hold, especially for the high-T_C superconductors where in some cases more than one energy gap are present. For NbN, the equation already gives a good estimation of the energy gap for thick films or bulk samples. The exact ratio $2\Delta(0)/k_B T_C$, however, depends once again on the film thickness and the substrate material used, which leads to a number of reported values. For example, reports on 250 nm thick films on silicon give an energy gap $2\Delta(0) = 3.9 k_B T_c$ [82]. Tunnel spectroscopy measurements of thinner films (10 nm) on sapphire indicate a stronger coupling of $2\Delta(0) = 4.15 k_B T_C$ [5]. Far-infrared transmission experiments come to conclusions anywhere between BCS gap [83] and a strong coupling of $2\Delta(0) = 4.4 k_B T_C$ [84], depending on film thickness and substrate.

This rises the assumption, that in the NbN films not only T_C itself is changed with the chemical composition, but also its relation to the energy gap. So far, no previous reports of the dependence of $2\Delta(0)/k_B T_C$ on the NbN stoichiometry exists. Additionally, it is beneficial to measure the energy gap directly at the operation temperature 4.2 K of typical SNSPD for the evaluation of the minimum photon energy for the creation of a hot spot, as described in equation 2.6. Direct measurements by THz transmission spectroscopy in the range 0.1-1.2 THz is conducted on a series of unpatterned 4 nm thick NbN films deposited on sapphire substrates. The measurements were performed at the 1. Physikalisches Insitut at the University of Stuttgart. The results described in the following were partly published in [85].

The setup consists of a set of tunable monochromatic THz sources and a Mach-Zehnder interferometer. To measure the transmission perpendicular to the sample surface, the radiation is chopped and polarized and then directly detected after passing through the sample in a cryostat. The polarization is set along the main axes of the R-plane sapphire substrate to avoid the birefringence of the material [86]. The phase is obtained in the Mach-Zehnder configuration, i.e. the beam is split and only one path send through the sample, and then recombined by a second beam

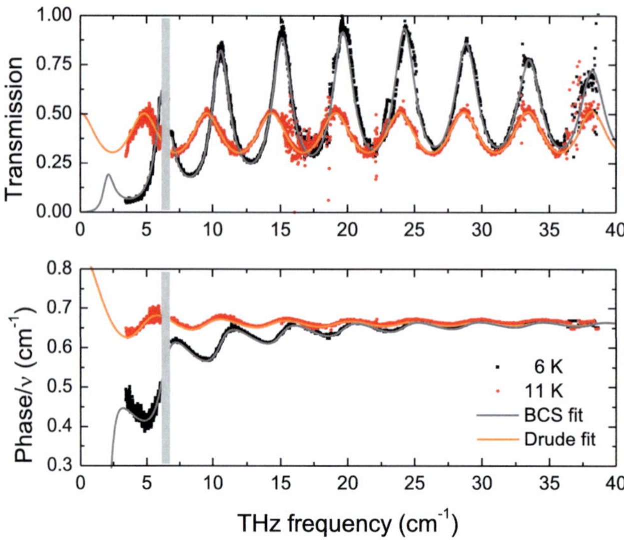

Figure 3.13: Transmission (top) and phase (bottom) on frequency for a 4 nm thick NbN film on sapphire. The gap in the measurements between 6.1 and 6.9 cm^{-1} is experimentally inaccessible. The red data are taken at 11 K, which is above T_C, whereas the black data were taken at 6 K, which is well in the superconducting state. The solid lines show fits of the Drude model for the normal state and the BCS model for the superconducting state to the data.

splitter with the second, undisturbed path. After passing through the analyzer the radiation is detected by a Golay cell. The transmission and phase of the signal show typical interference patterns that are Fabry-Perot resonances of the sample layer system. The measurements are shown in Fig. 3.13 exemplary for one sample at one temperature in the normal state (red) and one in the superconducting state (black).

The data is analyzed with the help of a model of the two-layer system. The optical properties of the substrate are determined beforehand by a measurement of a bare sapphire substrate. The optical conductivity of the thin film is either described by the Drude model for measurements at $T > T_C$ or by the BCS theory for $T < T_C$. From the optical conductivity, the transmission and phase can be calculated in the dirty limit with the Mattis-Bardeen formalism [87]. The results depend only on frequency ω, temperature T and energy gap 2Δ. For each sample and each fixed temperature, the measured spectra of transmission and phase are fit simultaneously by the model, with 2Δ as fit parameter. The fits are shown as solid lines in Fig. 3.13 for the presented sample. The NbN films can be well described by these models and the energy gap is extracted from the fits with a high accuracy.

As a result, two values (one for each polarization) of the energy gap 2Δ are obtained for each temperature of the measurements. Figure 3.14 left shows the results of the measurements for four 4 nm thick films deposited at different sputter currents. On first observation, the depen-

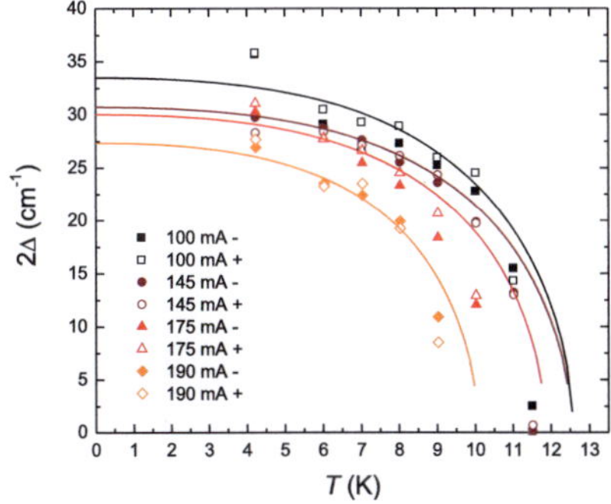
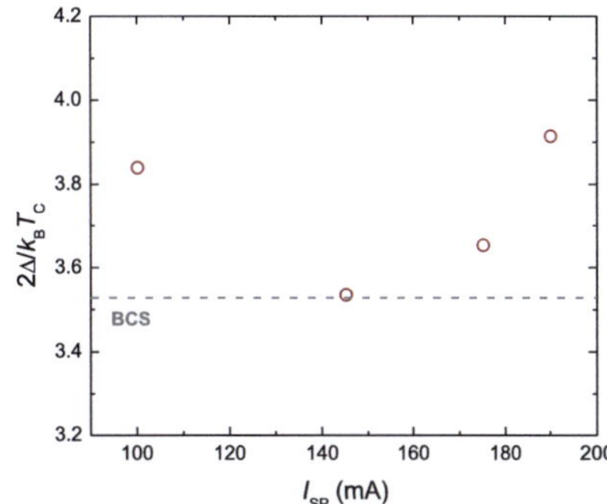

Figure 3.14: Left: Temperature-dependence of energy gap from fits to THz transmission spectra for 4 nm thick NbN films deposited at different sputter currents as indicated in the legend. Solid symbols are measured with negative polarization, open symbols with positive polarization. The solid lines are fits on eq. 3.7 to the data with fit parameter $\Delta(0)$. Right: $2\Delta/k_B T_C$ extracted from the measurements over the sputter current I_{SP} at which the film was deposited. The dashed line marks the BCS value.

dences seem to scale with the T_C values of the films. As an approximation for the temperature-dependence of the energy gap, we use the expression

$$\Delta(T) = \Delta(0) \left[1 - \left(\frac{T}{T_C}\right)^2\right]^{\frac{1}{2}} \left[1 + \left(\frac{T}{T_C}\right)^2\right]^{\frac{3}{10}}. \tag{3.7}$$

The solid lines show fits of eq. 3.7 to the data, with the zero temperature gap $\Delta(0)$ as fit parameter. Except for values close to T_C, the dependence describes the data well. The discrepancy in T_C might be due to slight degradation during transport of the films, as for the fit the values obtained in the pre-characterization of the films were used. However, a change in T_C in the fit has only small influence on the $\Delta(0)$ value.

Finally, for each film the ratio $2\Delta(0)/k_B T_C$ is calculated and shown in Fig. 3.14 (right) over the sputter current at which the film was deposited. As a reminder, the films deposited at low I_{SP} have an increased nitrogen content, and at high I_{SP} an increased niobium content. The $2\Delta(0)/k_B T_C$ values seem to follow roughly the inverse dependence of T_C. A minimum is found at 145 mA, where the maximum T_C values are achieved. Here, the value 3.536 is almost the BCS value of 3.528. For both higher and lower I_{SP}, the values increase, indicating a more strongly coupled superconductor. However, in total the dependence is not very strong. For 190 mA the value is 3.914, which is only about 10% larger than the minimum.

The energy gap enters eq. 2.6 quadratically. Even the small corrections for the material dependence found here should thus be included in the analysis. However, only the gap value at the measurement temperature is required, which is typically 4.2 K for SNSPD characterizations. For the analysis presented in chapter 6, we therefore use the energy gap values directly from the measurement point at this temperature. For the standard composition of $I_{SP} = 145\,\text{mA}$, for example, the energy gap value is $2\Delta(4.2\,\text{K}) = 3.7\,\text{meV}$.

3.5 Dependence of Optical properties on deposition current

All of the parameters evaluated so far in this section influence on the intrinsic detection efficiency of the material. However, from an application point of view, a high intrinsic detection efficiency alone is not sufficient. The probability of the absorption of a photon incident of the film's surface is given by the absorptance of the film. The overall detection efficiency can never exceed the absorptance unless the light is coupled into the film by other means. Since the material variation changes the electronic state, it is expected that the absorptance of the film will also vary. The case has been already studied in great detail for NbN films deposited with the standard composition, i.e. the conditions at which the highest T_C values are obtained [19]. Here it has been found that the absorptance changes with the thickness of the films, as would be expected since the shorter optical path reduces the chance of absorption of a photon. However, it was found also that the absorption coefficient itself and its dependence on the wavelength change with the film thickness, which was attributed to the change in material properties.

In this section, the optical properties of NbN films with the same thickness, but varying chemical composition, are investigated with ellipsometric measurements. It is not the aim to derive a full wavelength-dependence of the absorptance for each film, since in the following sections mainly the intrinsic detection efficiency is discussed. However, for a complete assessment of the suitability in a detector application, the absorptance is evaluated at two selected wavelengths: 850 nm to reflect the detection of optical or near-infrared photons and 1550 nm as a key wavelength for infrared applications.

3.5.1 Ellipsometry

In ellipsometry, information on a sample's layer system is obtained by reflecting a laser beam off the sample surface. It is often employed to determine the thickness of a sample or even several layers with very high accuracy when the optical parameters of the materials involved are known. If on the other hand the thickness is known, it can be used in the same way to determine the complex dielectric function $\tilde{\varepsilon}$. There exist many different ellipsometric setup variants. The measurements described in the following sections were made with a VASE ellipsometer from J. A. Woollam, which is briefly described: It consists of two separately rotating arms and the sample holder. On one arm, a monochromator source and a polarizer allow for the selection of a light beam with known wavelength and known linear polarization. The light is incident on the sample surface under a certain angle θ, where the polarizations perpendicular (s-polarized)

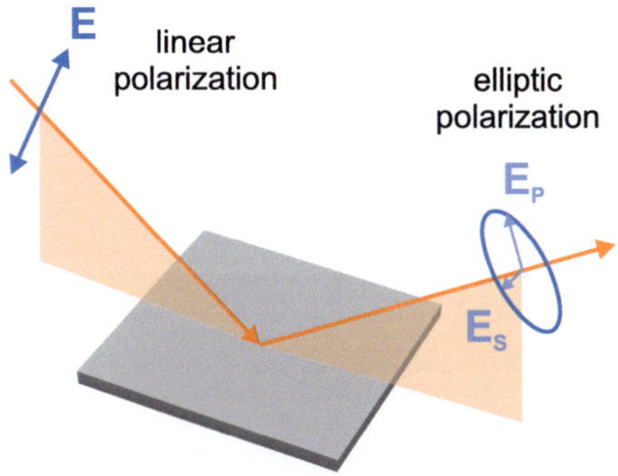

Figure 3.15: Schematic of ellipsometry measurement configuration. Linearly polarized light is reflected from the sample surface and the interaction leads to elliptically polarized outgoing light. The polarization change is detected with a rotating analyzer and a photodetector.

and parallel (p-polarized) to the plane of incidence are reflected differently (see Fig. 3.15). The reflected light is thus polarized elliptically with a phase difference Δ between the two polarization axes and an angle Ψ between the long axis of the ellipse and the s-polarization. The other arm is equipped with a rotating analyzer and a four-quadrant photodetector that measures the ratio of the complex amplitudes of the s-polarized (r_s) and p-polarized (r_p) reflected light:

$$\frac{r_p}{r_s} = \tan(\Psi)e^{i\Delta}. \tag{3.8}$$

Both arms can be moved to change the reflection angle θ. For maximum sensitivity of the ellipsometric measurements, an angle close to the Brewster angle should be employed. For typical NbN films, this angle is roughly $58°$. The measurements were therefore taken at an angle of $60°$. Since the sample should be homogeneous over the area of the light spot, which is in the order of 1 mm in diameter, the ellipsometric measurements are made on unstructured NbN films that were deposited on polished sapphire substrates. The back-side of the substrate is unpolished, so that the light scatters diffusively and does not influence on the reflection measurement. For calibration of the system, a bare substrate was used. The measurements were taken in the wavelength region 400-1600 nm.

3.5.2 NbN Multi-layer system

The measured parameters Ψ and Δ are directly connected to the complex dielectric function $\tilde{\varepsilon} = \varepsilon_1 + i\varepsilon_2$ of the material (and thus to the index of refraction $\tilde{n} = n + ik$ via $\tilde{\varepsilon} = (\tilde{n})^2$) and can be recalculated for a single layer of material with well-known thickness. However, the determination of the exact thickness with high accuracy is in itself challenging for NbN films below 5 nm. Also, films can only be measured by ellipsometry when the transmitted signal is high enough, so the film thickness may not be too high. The analysis is further complicated by the fact that the considered films consist of multiple layers as discussed in section 3.1.2. In this case, a model for the layer system has to be created from which the values of Ψ and Δ are computed and iteratively fitted to the measurement data. A detailed descritpion of the thickness measurement and the development of the layer model for the NbN films is given in [88]. The result is briefly described in the following:

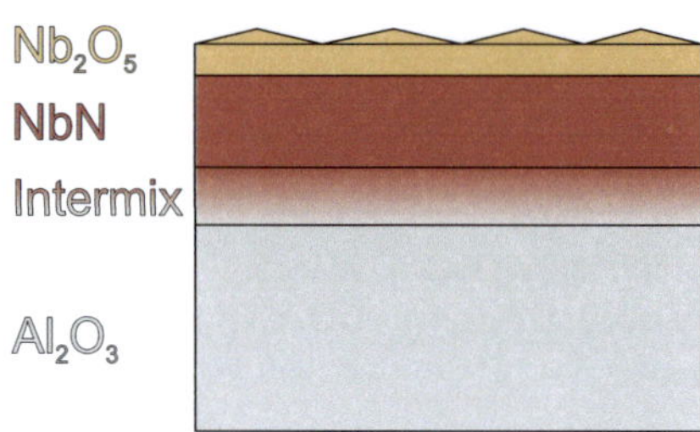

Figure 3.16: Layer system used for the modeling of the ellipsometric measurement. The basis is the semi-infinite sapphire substrate. At the boundary to the NbN film, a gradual intermix layer is formed. On top of the NbN itself is an oxide zone, and an additional surface roughness layer accounts for non-uniformity in the surface.

First, measurements of several bare substrates are averaged and the optical parameters extracted with a simple model of a single layer with infinite thickness. These optical parameters are then used for the substrate as a basis of the model. The main layers are the substrate layer of infinite thickness and the NbN layer on top, of which the optical properties are fitting parameters (under conservation of the Kramers-Kronig relation). Between the NbN layer and the substrate, the interim region is described by a 1 nm thick intermix-layer in which the optical parameters are gradually changed from one material to the other. On the top of the NbN, a 0.7 nm thick layer of Nb_2O_5 represents the surface oxide layer. Although the optical parameters for this layer are not very well known, a change in these results in almost unmeasurable differences in the model prediction, so the standard values from the database suffice. Lastly, on the top of the oxide layer a surface roughness layer accounts for the non-ideal flatness of the sample surface. Figure 3.16 shows schematically the layer structure of the model used for the iterative fitting. The thickness of the intermix layer and the oxide layer correspond to the evaluations of TEM-measurements [19] and are subtracted from the total film thickness to give the thickness of the NbN layer in the model.

The remaining free parameter in the model is the total film thickness. It influences into the calculations strongly and has to be evaluated with high accuracy. For each film that is measured with the ellipsometer, a second one was made in the same deposition run for thickness measurements. The second film was structured by photo-lithography and reactive ion etching in a CF_4/O_2 plasma. The etching process has a high selectivity, so that etching into the substrate can be neglected. On these structures, the film thickness is measured with a profilometer. To verify the profilometric measurements, one of the NbN films was additionally measured with a high resolution atomic force microscope at the Institute for Nanotechnology at the KIT. The film thickness was previously measured with the profilometer to 3.63 nm. The measurement with the AFM yielded a value of 3.69 nm, which supports the confidence in our thickness evaluation.

3.5.3 Absorptance of NbN thin films

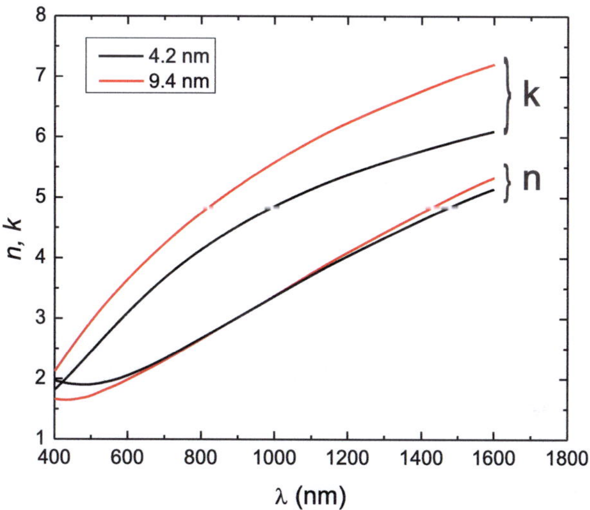

Figure 3.17: Spectra of the real (n) and imaginary (k) component of the refractive index for two NbN films of different thickness, deposited at 145 mA. For the thicker film (red lines), the absorption-related k is higher and steeper than for the thinner film (black lines).

From the least mean square model fit to the measurement data and with the known layer thicknesses, the dielectric function of the NbN film is extracted. Figure 3.17 shows the dispersion relation of the real and imaginary part of the refractive index exemplary for two NbN films. The films are both deposited at the standard 145 mA deposition current but have different film thicknesses. The refractive index n is below 2 for low wavelengths, displaying a minimum in the optical region depending on the film thickness. Towards the infrared range however, it rises steadily up to about $n = 5$ for 1600 nm. The imaginary part k is higher for the thicker film in the whole spectral region. k is closely related to the absorbance of the film, confirming that the thicker film shows an overall higher absorbance than the thinner one. The fact that k rises monotonically with the wavelength was observed for all studied NbN samples.

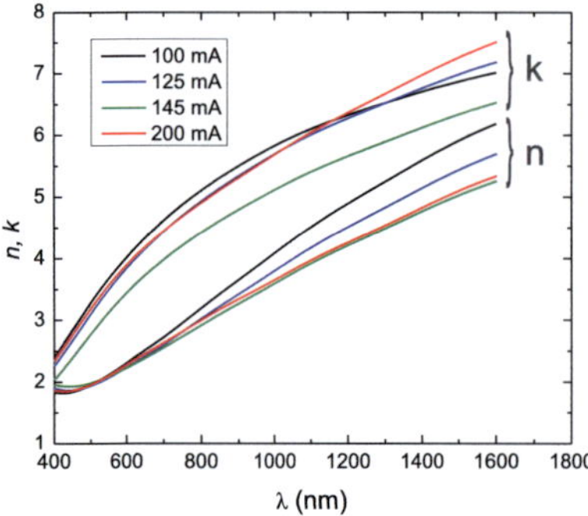
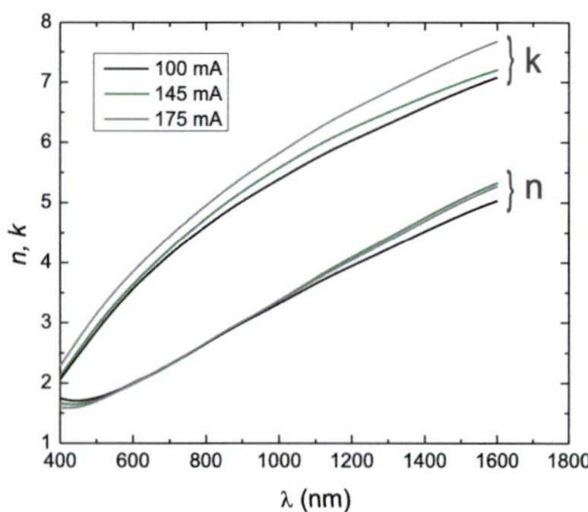

Figure 3.18: Spectra of the real (n) and imaginary (k) component of the refractive index for two NbN films of constant thickness 6 nm (left graph) and 10 nm (right graph). The films are deposited at varying sputter currents as indicated in the legend.

To evaluate the influence of the chemical composition of the NbN films on the dielectric function, two series of 6-7 films have been deposited with sputter currents ranging from $I_{SP} = 100 - 225$ nm. To measure directly on the 4 nm thick films that were used in the detector measurements would have given a high inaccuracy for two reasons: First, the lower absorption of a thinner film leads to a smaller signal and thus a larger error margin at the ellipsometric measurement itself. Secondly, the thickness measurement with the profilometer is more difficult for thinner films, increasing the inaccuracy further through the model fit. As a compromise that is still considered to be in the thin film regime, one series was deposited with thicknesses of $d = 6.3 \pm 0.9$ nm and another one with $d = 9.9 \pm 1.7$ nm. The dielectric functions extracted with the ellipsometric measurements are shown in Fig. 3.18 for some of the samples. The variance in the exact thickness leads to an offset in the k-values along the y-axis, but the films all show similar behavior. However, the change in the dispersion can be clearly seen: The k-values rise stronger with the wavelength for the films with the higher sputter current. The refractive indices, on the other hand, show the opposite tendency.

As a rough estimation for the absorptance, we determine the absorption coefficient

$$\alpha = \frac{4\pi}{\lambda} k, \tag{3.9}$$

from the imaginary part of the refractive index k and calculate the absorptance A via the Lambert-Beer law

$$A = 1 - \exp(-\alpha d). \tag{3.10}$$

The resulting dependencies on the sputter current of the films are compared for $\lambda = 850$ nm and $\lambda = 1550$ nm in fig. 3.19. There is a tendency towards higher absorptance values for the films with higher sputter currents that is more strongly pronounced in the 10 nm thick series. Though

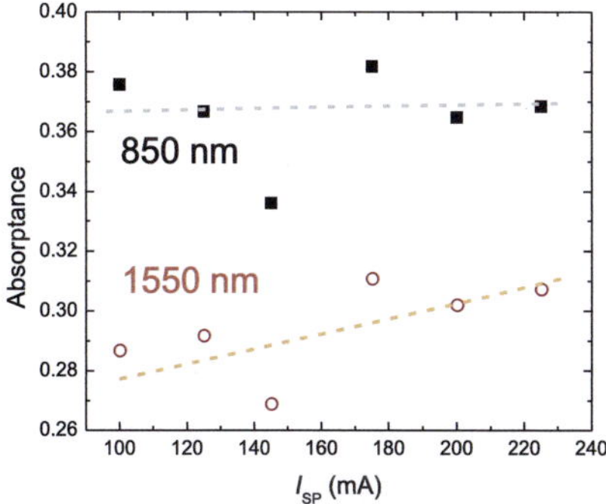
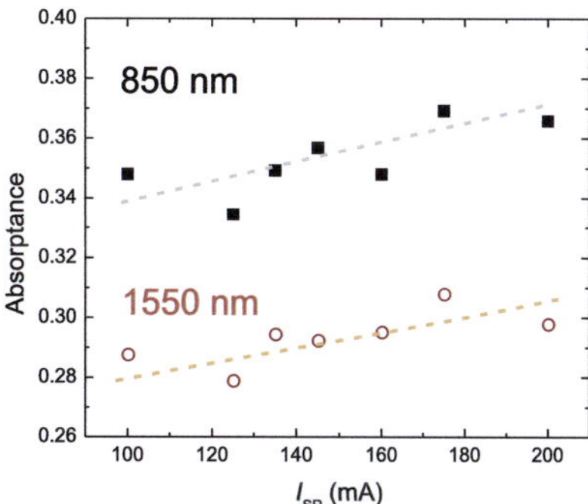

Figure 3.19: Absorbance estimated from the imaginary part of the refractive index, k, for NbN films of 6 nm thickness (left graph) and 10 nm thickness (right graph), deposited at varying sputter currents I_{SP}. Two selected wavelengths are shown: 850 nm (black squares) and 1550 nm (red circles).

the absorption is dependent on the wavelength, this tendency remains. The total absorptance values are in all cases lower for the higher wavelength due to the larger λ/d ratio, that reduces the probability of photon absorption. However, the overall dependence of the absorptance on the film composition is not very strong, varying roughly between $A = 33 - 38\%$ for $\lambda = 850$ nm and $A = 27 - 32\%$ for $\lambda = 1550$ nm.

In summary, it was found by means of ellipsometric measurements that the change of the material composition influences on the dispersion relation of the dielectric function of the NbN material. The resulting absorptance of the NbN thin film can be estimated and is in the order of 0.3. The exact value depends on the film thickness and the considered wavelength. For all NbN compositions, the absorptance values are high enough to allow for a feasible detector application. However, the material appears to become more optically dense as the Niobium content is increased, leading to more favourable absorption conditions.

3.6 Summary

This chapter presented a detailed investigation of growth and material parameters of NbN thin films which were deposited by reactive magnetron sputtering on heated sapphire substrates. The films are uniformly granular and the superconducting and resistive properties are strongly influenced by the grain boundaries. Due to the different growth conditions between initial and upper layers as well as the oxidation of the surface layer, the films show effectively a multilayer structure. Their properties, especially for $d < 10$ nm, are very sensitive to the deposition conditions, the substrate material, and the film thickness. Deposition conditions were evaluated for which the chemical composition of the films can be varied controllably and reproducibly via the variation of the sputter current I_{SP}. Especially for thin films, the reduction of the sputter current leads to a transition from metal to almost isolator: The resistivity increases, the temperature coefficient of the resistivity changes from positive to negative and the electron diffusion coefficient is reduced.

Superconducting NbN forms for sputter currents in-between those extremes. The optimal value of T_C is found at $I_{SP} = 145\,\mathrm{mA}$, independent of film thickness. Going to either higher or lower nitrogen content reduces T_C, but the dependence is steeper for reduced nitrogen content. The superconducting energy gap does not directly follow this dependence, because the conversion factor $2\Delta/k_B T_C$ itself rises for the non-standard chemical compositions, indicating a more strongly coupled superconductivity. However, the change is within 10% of the standard value, so T_C remains to be a good first estimate for the energy gap. For the case of 4 nm thick NbN films, the energy gap at zero temperature displays the highest value for the film with the highest nitrogen content and reduces with increased I_{SP}.

The critical current density also displays an optimum, but at higher I_{SP} than the critical temperature. Also, the optimum varies with the film thickness of the samples. Apart from this, the same strong reduction with increasing I_{SP} as for T_C is observed. However, j_C in a nanowire can significantly differ from values obtained on wide bridges, so conclusions on SNSPD performance are limited. Influences of the wire geometry on j_C are investigated in more detail in chapter 5.

For the realization of SNSPD with improved spectral bandwidth, the material properties should minimize the required photon energy for the hot-spot formation given by eq. 2.6. For a rough estimation, the current dependent term is disregarded and the so far unknown parameters ς, τ as well as the geometrical factors w, d are assumed to remain constant. To a first expectation the spectral bandwidth should profit from a reduced nitrogen content, because the energy gap is then reduced. However, those films also have a lower resistivity, which enters the equation inversely via the electronic density of states: $N_0 \propto (\rho D)^{-1}$. To achieve low E_{ph} values, the resistivity of the material should thus be preferably high, which would require deposition at low I_{SP}. The electron diffusion coefficient D remains with a dependence $E_{ph} \propto D^{-0.5}$. This gives once again a slight preference for the films with high D values which are deposited at high I_{SP}. Due to this opposite requirements on the inter-dependent material parameters, no clear preference for the NbN chemical composition is apparent at this point. A more detailed discussion including a quantitative evaluation, the current-depending term and a direct comparison with experimentally obtained E_{ph} will be given in section 6.2.

An additional preference to films deposited at high I_{SP} is given for applications where the photon is coupled into the NbN film by absorption. The films with increased Nb content appear to be more optically dense as would be expected from the increased number of charge carriers. The absorptance of a NbN thin film with typical thickness for SNSPD application in the optical and near-infrared range was evaluated to be about 30%-35%. The value depends on the chemical composition and the thickness of the NbN film as well as the considered wavelength. Especially for long wavelengths, the films deposited at high I_{SP} provide improved absorbance values.

4 Energy relaxation of quasiparticles in superconducting thin films

In a superconducting radiation detector, the dynamics of the detector response is defined by the relaxation processes of the energy that is absorbed by the quasiparticles in the superconducting film. Since the first photo-response measurement on superconductor films [89], many experiments on various superconducting materials were conducted using optical pulses [90, 91, 92, 93] and amplitude modulated THz radiation [90, 94, 95].

In the case of an SNSPD, the absorbed photon leads to a strong initial excitation of the quasiparticles and an almost instantaneous emergence of the normalconducting zone that causes the voltage pulse. Until the nanowire returns to the superconducting state, no new detection event can take place. The minimum time required for the back-switching is given by the intrinsic energy distribution processes that describe the time evolution of the hot-spot. The rates of this processes define the material limit for the minimum detector dead time that can be achieved. They can be well described in the framework of the Two-Temperature model, which is briefly reviewed in the first section of this chapter.

To identify the main contributing factors to the relaxation time of an SNSPD pulse, the energy relaxation processes taking place in a superconducting NbN thin film are studied with a frequency-domain measurement technique. The films are excited by modulated laser light with low intensity, which causes small periodic deviation from the equilibrium state. The operation conditions exclude contributions from vortex movement that were observed in experiments with sub-THz excitation [96, 97]. The response spectrum to varying modulation frequencies allows the extraction of the involved time constants with much higher precision than in the time domain. The principles of the measurement technique and the experimental setup are described in section 4.2. The dependence of the energy relaxation time on the NbN film thickness and on the material composition is investigated in two measurement series. The results are discussed in the framework of the Two-Temperature model to identify the limits of the maximum achievable SNSPD count rates.

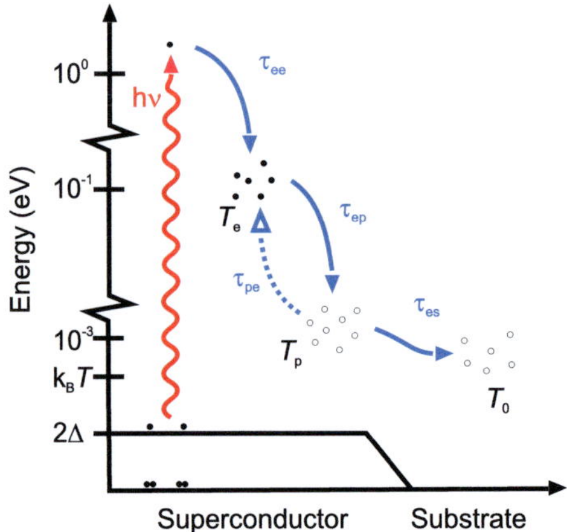

Figure 4.1: Schematic illustration of the energy relaxation processes in a superconducting thin film. After absorption of the photon (red), a quasiparticle is elevated to high energies. By electron-electron scattering with a rate τ_{ee}^{-1}, its energy is distributed to the electron subsystem, which can be assigned an effective temperature T_e. By electron-phonon interaction (τ_{ep} and τ_{pe}) the energy cascades further down to the phonon subsystem at energies T_p until the phonons diffuse out of the superconducting film to the substrate at bath temperature T_0.

4.1 Two-temperature model

An optical or infrared photon incident on a superconducting film is absorbed by one electron of a Cooper pair, which breaks up the pair into quasiparticles. Since the photon energies in the range of eV are much larger than the energy gap, which is in the meV range, the highly excited quasiparticle is then breaking additional Cooper pairs in an avalanche-like process which leads to a change of the Cooper pair density. The processes of the energy cascade are schematically illustrated in Fig. 4.1. After the initial absorption of a photon with the energy $h\nu$, the energy is redistributed within the electron subsystem (filled circles) by inelastic electron-electron scattering (or at high energies by scattering with optical phonons) with a time-constant τ_{ee}. Those processes are very fast in disordered superconductors, so that $\tau_{ee} < 1$ ps can be assumed, and lead to an accumulation of quasiparticle excitations just above the energy gap. After the energy of the quasiparticles is sufficiently reduced, the cross-section for inelastic electron-phonon scattering becomes significant and the energy is transfered to the phonon subsystem (open circles) with a characteristic time τ_{ep}, and can also be scattered back to the electrons with the time constant τ_{pe}. If this backscattering is much slower ($\tau_{ep} \ll \tau_{pe}$), the electron and phonon subsystem are almost decoupled and their energy distribution can be described by an effective temperature T_e and T_p. Finally, the phonons diffuse out of the thin film into the substrate with the escape time τ_{es} and relax to the bath temperature T_0.

In the case of $T \approx T_C$, when Joule heating is neglected and only small deviations from equilibrium are considered, the general coupled differential equations [98] describing the time evolution of the two temperatures can be linearized [92]:

$$
\begin{aligned}
C_e \frac{dT_e}{dt} &= \frac{\alpha P_{in}(t)}{V} - \frac{C_e}{\tau_{ep}}(T_e - T_p) \\
C_p \frac{dT_p}{dt} &= \frac{C_p}{\tau_{pe}}(T_e - T_p) - \frac{C_p}{\tau_{es}}(T_p - T_0),
\end{aligned}
\tag{4.1}
$$

where C_e and C_p are the electron and phonon specific heat, α is the absorption coefficient, V the device volume and $P_{in}(t)$ the incident radiation power. In the case of a periodic excitation with the modulation frequency ω, $P_{in}(t) = P_0 \cos(\omega t)$, the electron temperature change ΔT_e can be calculated numerically from the coupled differential equations (see Perrin and Vanneste [99]). With some additional assumptions the equations can also be solved analytically to a good approximation which gives the change in the electron temperature ΔT_e in dependence of the modulation frequency ω [99]:

$$
\Delta T_e = \frac{\alpha P_0 \tau_{es}}{(C_e + C_p)V} \sqrt{\frac{1 + \left(\omega \tau_{ep}\left(1 + \frac{C_p}{C_e}\right)\right)^2}{(1 + (\omega \tau_{es})^2)(1 + (\omega \tau_{ep})^2)}}.
\tag{4.2}
$$

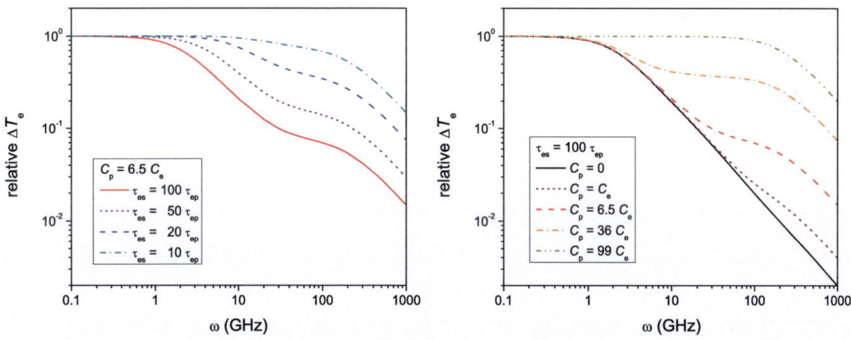

Figure 4.2: Frequency spectra of the change in electronic temperature for periodic excitation with frequency ω described by eq. 4.2. All curves are normalized to one for $\omega \to 0$ to show the relative dependence. The electron-phonon interaction time τ_{ep} is arbitrarily set to 5 ps. The left graph shows curves for various ratios of the time constants τ_{es}/τ_{ep}, the left graph for various ratios of the heat capacities C_p/C_e.

To illustrate the general dependence of ΔT_e given by eq. 4.2, the square root part of the equation is plotted in Fig. 4.2 for an arbitrary value of $\tau_{ep} = 5$ ps. The left graph shows the influence of the ratio τ_{es}/τ_{ep} for a fixed value $C_p/C_e = 6.5$. The spectra are in all cases constant at low frequencies, meaning that the change in the electron temperature ΔT_e will completely follow the excitation modulation. As ω approaches the inverse value of τ_{es}, the dependence shows a roll-off and decreases continuously, because the phonon escape into the substrate limits the relaxation of the excitation and the response amplitude is reduced. If τ_{es} is significantly larger than the electron-phonon interaction time τ_{ep}, a second plateau emerges in the dependence. Here, only the electron subsystem follows the excitation and thus yields a smaller temperature change ΔT_e. The second plateau may then extend up to higher frequencies that correspond to the inverse value of τ_{ep}, after which it will finally roll off again.

However, the exact frequencies and the shape of those roll-offs are not governed by the electron and phonon time constants alone, but also by the ratio of their specific heat capacities C_p/C_e. This influence is illustrated in the rigth graph of Fig. 4.2. For $C_e \ll C_p$ (yellow curve), the two systems can be treated as decoupled and only the roll-off of the electron system can be observed. If, on the other hand, the heat capacities are almost equal, the electron and phonon system are so strongly coupled that they almost act as one system and only one combined roll-off can be observed (brown curve). Just in the intermediate case, where C_p is clearly larger but not exceedingly larger than C_e, (and also $\tau_{ep} \ll \tau_{es}$) a clear double roll-off can be observed in $\Delta T_e(\omega)$. These conditions are given for example in YBCO, which was therefore studied extensively with optical excitation experiments. Here the ratio of the heat capacities is $C_p/C_e = 36$ [93] (orange curve). With an electron-phonon interaction time τ_{ep} in the picosecond range, the material can be used in the detection regime on the second plateau with smaller signal amplitudes ΔT_e, but allowing operation at very high frequencies and thus very good time resolution [100]. For NbN, the situation is different. Here, $C_p/C_e \approx 6.5$ [101] (red curve). Also the time constants are, depending on the film thickness considered, roughly in the range $\tau_{es} = 10\tau_{ep}$, so that a second plateau should be barely resolvable. Instead, a combined roll-off is expected at a frequency corresponding to an overall energy relaxation time τ_ε [102], where

$$\tau_\varepsilon = \tau_{ep} + \left(1 + \frac{C_e}{C_p}\right)\tau_{es}. \tag{4.3}$$

This energy time constant marks the intrinsic material limit for the relaxation of an SNSPD pulse and therefore the ultimate limit of any achievable count rate.

4.2 Frequency-domain measurement technique

To extract the energy relaxation time from the excitation by optical radiation, the response of the superconductor is measured in the frequency domain by application of discrete modulation frequencies. The device is operated at a temperature on the resistive transition, where a change in the electron temperature $\Delta T_e(f)$ incurred by the radiation is reflected by a change in the device resistance. The voltage $\Delta V(f)$ across the biased device is measured at the same frequency as the excitation frequency f with a spectrum analyzer.

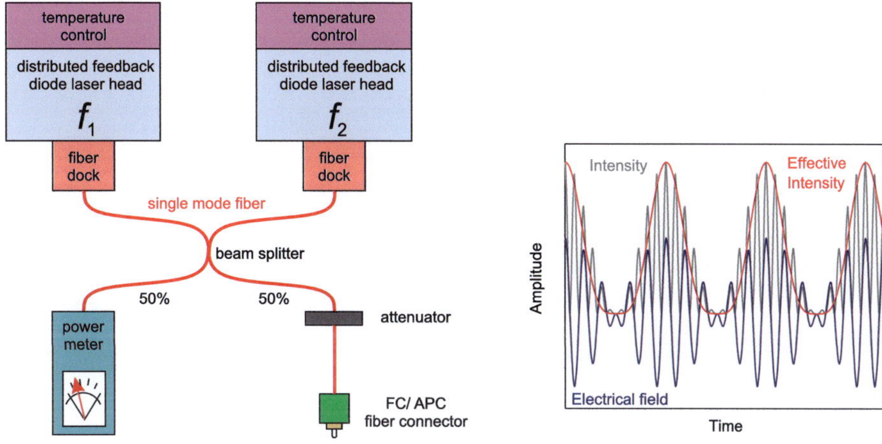

Figure 4.3: Schematic of the generation of modulated radiation by two cw lasers with slightly detuned frequency. Left: Laser light at $\approx$850 nm is generated by two distributed feedback lasers, where the exact frequencies f_1 and f_2 can be tuned very accurately by the laser's temperature. The beams are coupled via fiber ports into single mode optical fibers that lead to a symmetric fiber splitter. Here both fields combine in a linear superposition, which causes a beating of radiation at the frequency $f_1 - f_2$. The modulated light can then be attenuated by optical density plates in a microbench before it is connected to the measurement setup. The second fiber port is used to control the optical output power. Right: The resulting modulated electric field E at the output port is illustrated by the blue line. The fast frequency component in the electric field and in the respective radiation intensity E^2 (gray) will not be discernible by the sample. An effective intensity (red) can be used to describe the modulated output power.

The modulation of the radiation power is realized by the superposition of two cw laser sources. The exact wavelengths of the diode lasers are temperature-controlled. This allows the fine-tuning of their wavelength around 850 nm with very high accuracy. Both lasers are coupled into the same single-mode optical fiber. The superposition of the electric fields leads to a beating at the sum and the difference of the original laser frequencies f_1 and f_2 as illustrated in Fig. 4.3. The high frequency modulation of the resulting laser intensity will be outside the scope of the spectrum and can be disregarded. The effective laser intensity incident on the sample (red line) corresponds to an on/ off-modulation of the radiation at $f = f_1 - f_2$. The modulation frequency f can in this way be changed continuously from several Hz up to 300 GHz. The total maximum output power is 30 mW and can be tuned down by neutral density attenuator plates on an optical minibench that is inserted between fibers. The optical power is calibrated with a power meter for absolute values and a fast GaAs photodiode to determine the modulation depth of 24%. Both the optical power and the modulation depth stay within 5% of variation when the modulation frequency is varied in the measurement range from 10 MHz up to 10 GHz. The amplitude-modulated laser power is passed by a multimode optical fiber into the experimental insert into a liquid helium transport dewar and ends about 1 mm above the sample surface. The resulting light spot is much larger than the typical sample area, ensuring a homogeneous irradiation of the device. At typical measurement conditions, the density of the modulated radiaiton power at the sample position amounts to $\Delta P = 210\,\mathrm{pW/\mu m^2}$.

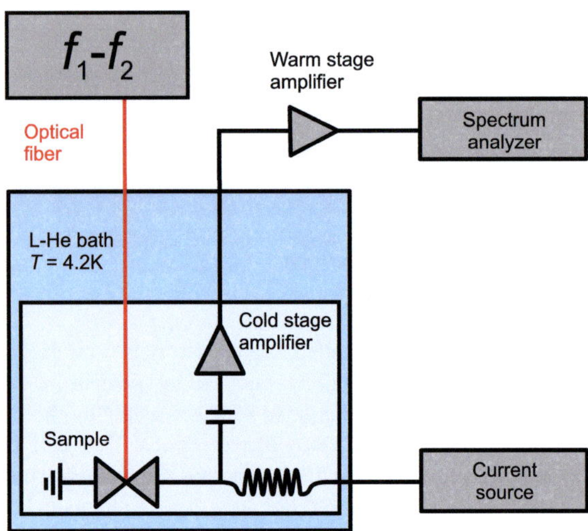

Figure 4.4: Schematic overview of the frequency-domain measurement setup. The modulated radiation created by the two cw lasers operating at frequencies f_1 and f_2 is fed into a multimode fiber leading into the transport dewar insert. The sample housing and the cold stage electronics are encased by variable contact gas (not shown) which allows adjustable coupling to the helium bath. The sample is biased by a low noise current source. The rf readout is separated by a bias Tee and pre-amplified by a cold stage amplifier before it is led out of the insert. After further warm stage amplification, the signal is sent to the spectrum analyzer.

A schematic of the experimental setup is shown in Fig. 4.4. The temperature of the sample is controlled by the contact gas pressure in the insert and a resistive heater in close vicinity of the sample. The sample is wire-bonded to an rf housing with an SMA connector. A bias-Tee separates the high frequency path from the DC bias current that is supplied by a low noise current source. The sample response is pre-amplified at cold temperatures by a two-stage low noise amplifier with 18 dB gain and a bandwidth of 12 GHz that was designed for low power consumption at cryogenic temperatures [103]. The signal is then led out of the insert by semi-rigid rf cables and after additional amplification sent to a spectrum analyzer. The readout path was calibrated at room temperature by replacing the sample housing with a signal generator and determining the transmission parameters, which showed a maximum dampening of 8 dB at 10 GHz. The cold stage amplifier was additionally characterized at liquid helium temperatures with a network analyzer and is set to a working point with optimal flatness below 0.5 dB in the considered frequency range. The measurement range of 10 MHz to 10 GHz was set by the bandwidth of the readout electronics with sufficiently constant transmission parameters.

The samples were patterned from NbN thin films on sapphire substrates that were deposited by reactive magnetron sputtering as described in chapter 3. The central part of the samples consists of a field of parallel stripes of 2 μm width and equal space between them. This design homogenizes the distribution of the bias current across the sample while retaining a large absorption area. The stripe number and their length were designed to realize a normal state resistance R_N of ≈ 150 Ω. The stripes are embedded into gold contact pads with a coplanar design layout that is matched to 50 Ω readout impedance (for more details see section 6.1.1). In addition to providing contacts, the gold serves to reflect radiation outside the detector area, so that the absorbed radiation is confined to the sample area. The central part of the samples were patterned by photolithography and subsequent reactive ion etching. After a second lithography step defining the contact layout, the sample surface was cleaned with an ion beam prior to the *in-situ* deposition of the gold contacts by magnetron sputtering. After lift-off, the samples were cut by a wafer saw into separate $3 \times 3 \, \mathrm{mm}^2$ sample chips. The chips were mounted into the optical measurement setup with silver paste to provide thermal contact between substrate and housing.

4.3 Investigation of the energy relaxation time of quasiparticles in NbN thin films

All samples were DC characterized in a separate setup prior to the optical measurements. In this section, the general behavior of the samples and the measurement analysis are first introduced on a typical sample. Following that, results on two sample series of varying thickness and with varying stoichiometry are discussed.

The samples are cooled down to an operation temperature on the resistive transition and are biased by a current I_B. When the electron temperature T_e is changed by incident radiation power, the resistance increases, leading to a respective change in the measured voltage

$$\Delta U = I_B \frac{\mathrm{d}R}{\mathrm{d}T} \Delta T_e. \tag{4.4}$$

A typical $R(T)$-curve of one sample is shown in Fig. 4.5. The derivative $\mathrm{d}R/\mathrm{d}T$ (brown) reaches the highest values at about one third of the normal state resistance. This point marks the optimal operation temperature for the measurements. At this operation temperature and a fixed modulation frequency of 150 MHz, the voltage change is then measured in dependence of the applied bias current I_B. The result is shown in Fig. 4.6 for a sample made from a 22 nm thick NbN film.

At small bias currents, the signal increases with increasing I_B nearly linearly as would be expected from eq. 4.4. As I_B is further increased, the signal reaches a saturation and then decreases.

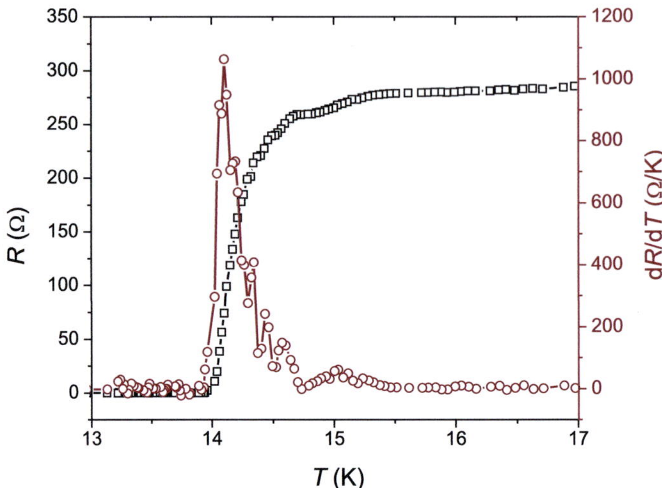

Figure 4.5: Superconducting transition $R(T)$ (black squares) and its derivative with respect to T (brown circles) for a typical NbN sample.

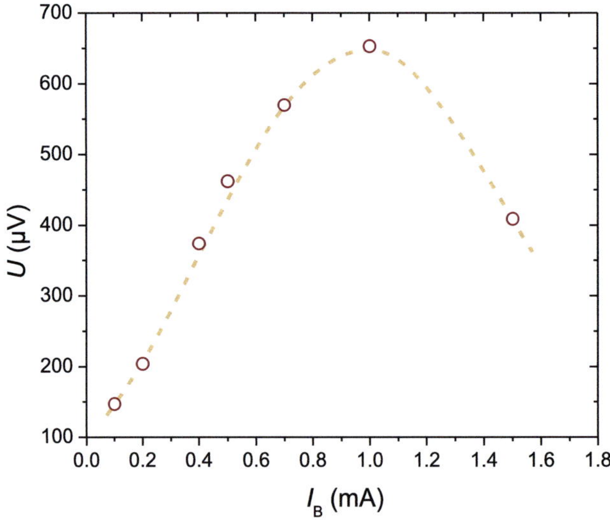

Figure 4.6: Signal output ΔU in dependence on the applied bias current I_B for a fixed modulation frequency of 150 MHz. The dashed line symbolizes the general behavior.

This is caused by the dissipated power that overheats the sample and causes a change in the working point. For further measurements, the operation was thus limited to the linear regime of the dependence.

Likewise, the optical power incident on the sample has a linear regime at low power and a saturation regime when the absorbed power changes the effective operation temperature. The optical power for the measurement was also restricted to the linear regime. For such operation parameters, the change in the electron temperature ΔT_e can be calculated from eq. 4.4 with the dR/dT-values taken from the previously measured $R(T)$ dependence and ΔU calculated with the help of the readout calibration. For the highest I_B and P_0 values used, $\Delta T_e = 125$ μK for the sample described above. This is significantly less than the excitation usually caused by pulse response measurements, where the electrons are heated by several Kelvins [104] and confirms the small signal regime of the measurements.

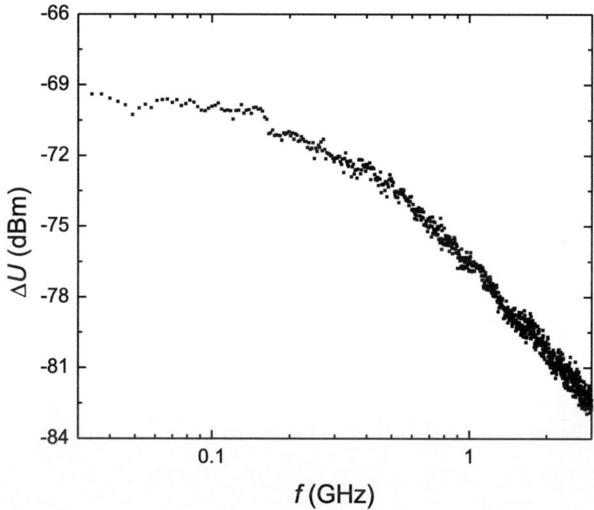

Figure 4.7: Spectral dependence of the voltage signal ΔU measured on a sample structured from a 8.6 nm thick NbN film. The dependence is flat for low f and then decreases gradually. Above $f = 2$ GHz values are not shown because the signal runs into the noise level.

Once the correct operation parameters are found, the modulation frequency $f = \omega/2\pi$ is varied to determine the spectral dependence of ΔU. A typical spectrum is shown as an example in Fig. 4.7. As expected from the description in section 4.1, the signal is almost constant for low frequencies but shows a roll-off dependence as f is increased. The measurements extend up to 2 GHz before the singal is reduced to the noise level. No second plateau, but one combined roll-off is observed as expected for NbN. A change in I_B or T changes the amplitude of the response and thus shifts the spectrum up or down. However, the relative dependence remains unaffected as long as both parameters stay within their respective linear regimes.

To extract the energy relaxation time constant τ_ε, the spectral dependence is fitted with the expression

$$\Delta U(f) = \frac{\Delta U(0)}{\sqrt{1+(\frac{f}{f_\varepsilon})^2}},$$

(4.5)

with $f_\varepsilon = (2\pi\tau_\varepsilon)^{-1}$ and $\Delta U(0)$ as fit parameters. In the case of NbN thin film, this time constant is related to the intrinsic scattering time constants by eq. 4.3. The electron-phonon interaction τ_{ep} as well as the ratio of the heat capacities C_e/C_p are expected to be material properties, whereas the phonon escape rate τ_{es} depends also on the film thickness and the interface transparency between film and substrate.

4.3.1 Thickness-dependence of energy relaxation time in NbN films

To determine the influence of the characteristic phonon escape time τ_{es}, dependence of the energy relaxation time constant τ_ε on the film thickness is investigated for a series of samples structured from NbN thin films. The results were partly published in [105], together with similar investigations on YBaCuO thin films.

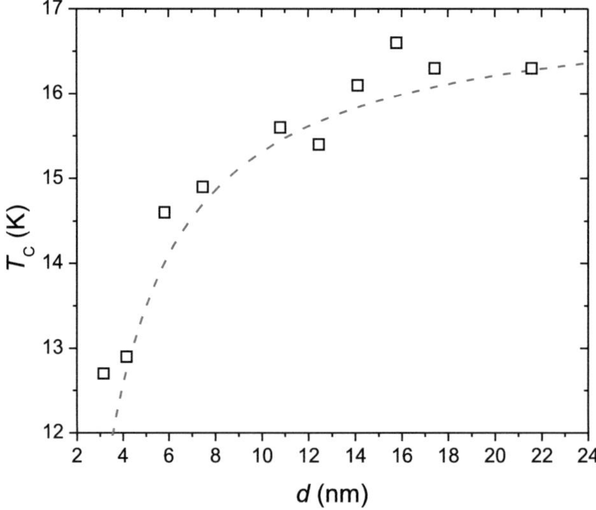

Figure 4.8: Dependence of critical temperature on the film thickness for the frequency-domain measurement samples structured from NbN. The dashed line shows the dependence typical for NbN thin films.

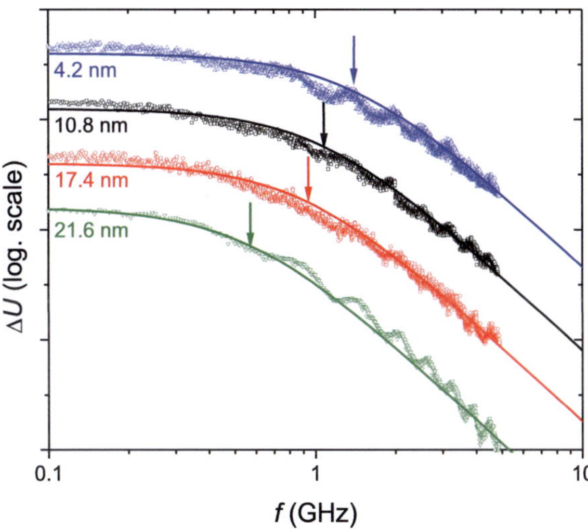

Figure 4.9: Dependence of response signal on modulation frequency f of four samples made from NbN films with different thickness as indicated in the graph. The spectra are shifted in the y-axis. The solid lines are fits of eq. 4.5 to the data; the arrows mark the roll-off frequency f_ε extracted as a parameter from the fit.

A series of NbN thin films was deposited at deposition conditions for optimal T_C on sapphire substrates. The thickness of the films was varied from 3.2 to 33.2 nm. The films were structured into samples for the frequency-domain measurements as described in the last section. After fabrication, the resistive transition was determined on all devices in a separate DC measurement setup. The T_C values extracted from these measurements range from 12.7 K for the thinnest film to about 16 K for the thick films and are shown in Fig. 4.8. The dependence on the thickness of the film reflects the general dependence found for the unstructured films (compare Fig. 3.3). The derivative of the resistance with respect to the temperature dR/dT showed for all samples a maximum at $\approx 0.3R_N$ with a value in the order of $1000\,\Omega/\text{K}$.

After characterization, the photo-response of the samples to 850 nm radiation was measured in the frequency domain. For the measurements, the samples were cooled to the temperature at which dR/dT reached its maximum value. The resulting spectral dependence of the signal is shown for four samples in Fig. 4.9. The spectra were shifted along the y-axis to be better discernible. All spectra show the expected plateau for low frequencies and the roll-off towards higher frequencies. However, the roll-off is shifted to lower frequencies as the film thickness of the sample increases. The remaining impedance mismatch between the sample and the readout circuit gives rise to a parasitic modulation of the signal, which oscillates with $\approx 500\,\text{MHz}$ and scales with the signal strength.

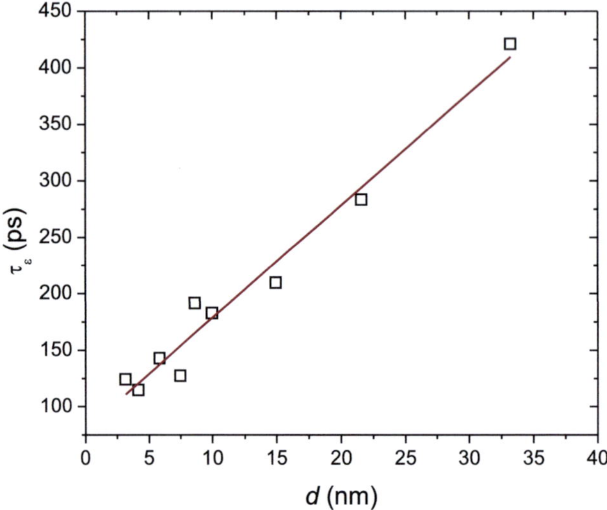

Figure 4.10: Energy relaxation time constant τ_ε in dependence of the NbN film thickness. The solid line is a linear fit to the data.

The least mean squares fit of eq. 4.5 to the data is shown by the solid lines. The respective fit parameter f_ε, indicated by the arrows, gradually reduces as the film thickness is increased. A change in optical power or bias current within the respective linear regimes only leads to a change in the amplitude, but results in the same f_ε with some statistical variation. The spectrum is therefore for each sample measured at several P_0 and I_B, and the average f_ε is taken to calculate the energy relaxation time $\tau_\varepsilon = (2\pi f_\varepsilon)^{-1}$. The resulting dependence on the NbN film thickness is shown in Fig. 4.10. For the thickest film with 33.2 nm, a relaxation time constant of 421 ps was determined. τ_ε decreases gradually with the thickness, down to 124 ps for the smallest thickness of 3.2 nm.

The electron-phonon interaction time is a material property and depends on energy and temperature. As a first approximation, we assume it to be independent of the sample thickness, as long as the material properties do not themselves change significantly with the thickness. The phonon escape, however, takes place only at the interface. The mean distance to the interface rises with the film thickness, so the phonons have to diffuse a longer way before they can leave the superconducting film. The probability that the phonon then passes the interface instead of being reflected can be describe by the phonon interface transparency α. The phonon escape time is thus proportional to the mean diffusion time d/v_S (where v_S is the phonon speed in the film) and inversely proportional to the interface transparency:

$$\tau_{es} = \frac{d}{\alpha v_S}. \tag{4.6}$$

Inserting this in eq. 4.3, an overall linear dependence of τ_ε on the film thickness is expected. Indeed, the linear dependence fits the data quite well, see solid line in Fig. 4.10. The slope of the fit is 9.9 ± 0.6 ps/nm. However, it should be noted that especially for the very thin films, the accuracy of the results is strongly limited by the neglected temperature-dependence. Since the measurement technique can only be employed at operation temperatures close to T_C, the strong reduction of T_C for the thin films (compare Fig. 4.8) leads to a significant change in the measurement temperature. This should influence on the heat capacities C_e and C_p, on the electron-phonon scattering time constant τ_{ep} as well as possibly on v_S and α.

The obtained values are somewhat higher but comparable to values determined by an electro-optical sampling method [101]: Here, for a 3.5 nm thick NbN film and a measurement temperature of 10.5 K, time constants of τ_{ep}=10 ps and τ_{es}=38 ps were measured, leading to a relaxation time constant τ_ε=53.7 ps.

4.3.2 Stoichiometry-dependent energy relaxation time in NbN films

To investigate the dependence of the relaxation time constant on the material, a second series of samples is made from NbN films deposited at varying sputter current I_{SP}. The films are made with the same thickness of ≈ 10 nm. The characteristics of the structured samples are compiled in table 4.1. The critical temperature T_C shows the dependence introduced in chapter 3, with a maximum at $I_{SP} = 145$ mA. The square resistance R_S decreases as the stoichiometry is shifted to increased niobium content at higher I_{SP}.

Table 4.1: Sample series for relaxation time measurements structured from NbN films with varying chemical composition. The table gives the sputter current I_{SP} of the film deposition, the film thickness d, the square resistance R_S, the critical temperature T_C as well as the cut-off frequency f_ε and energy relaxation time constant τ_ε extracted from the frequency domain measurements.

| I_{SP} | d | R_S | T_C | f_ε | τ_ε |
mA	nm	Ω	K	GHz	ps
100	9.6	122.4	14.34	0.986	161.5
125	8.2	132.8	14.76	1.075	148.1
135	9.1	132.5	14.89	0.933	170.6
145	9.4	89.2	15.15	0.803	198.3
160	10.0	80.9	15.01	0.801	198.7
175	10.0	74.3	15.10	0.840	189.6
200	11.6	84.6	14.06	0.835	193.0

The response spectrum is then measured for each sample in the frequency-domain setup at various P_0 and I_B. The cut-off frequency f_ε is determined from each spectrum by fitting eq. 4.5 to the data and then averaged. Three selected spectra from different samples are shown in Fig. 4.11. The film with the highest T_C at 145 mA shows the smallest f_ε, while the best value is achieved for the film with the increased nitrogen content. The values determined in this way and the calculated energy relaxation time constants τ_ε are listed in table 4.1.

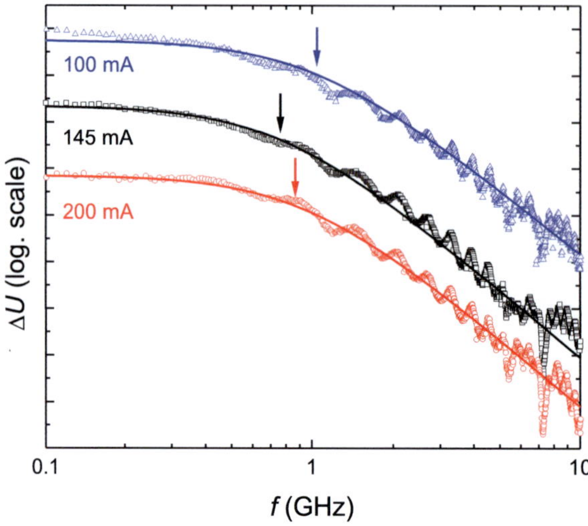

Figure 4.11: Dependence of response signal on modulation frequency f for three samples made from different NbN films. The sputter current of the film deposition is indicated in the graph. The spectra are shifted in the y-axis. The solid lines are fits of eq. 4.5 to the data; the arrows mark the roll-off frequency f_ε extracted as a parameter from the fit.

The variation of τ_ε with the sputter current shows values between a minimum of 148.1 ps to a maximum of 198.7 ps. The dependence is shown in Fig. 4.12. The scattering of the results is most likely due to the remaining variation of the sample thickness. The T_C-variation within the sample set is less than 1 K, so no significant temperature-dependence has to be expected. However, a general trend to higher τ_ε values for higher I_{SP} is observed (dashed line).

The electron-phonon scattering probability should generally be higher in materials that exhibit less order in their structure, because the defects act as scatter centers. Indeed in pure niobium films, reported relaxation time constants are in the nanosecond range [106], much larger than for the more disordered NbN. This tendency is reflected in the higher τ_ε values for the films with increased Nb content. This further confirms the more uniform structure of the films with increased Nb content.

To achieve minimum pulse relaxation times in nanowire detectors, this would favor the NbN films with reduced Nb content. However, the reduction of τ_ε of the film with $I_{SP} = 100$ mA with respect to the standard composition is only 18.6%. For all considered chemical compositions, the energy relaxation time constant is below 200 ps, which is low enough to allow fast detector operation. In typical SNSPD layouts, the relaxation time constant due to the nanowire's inductance has typically much higher values and thus remains the limiting factor for the maximum count rate.

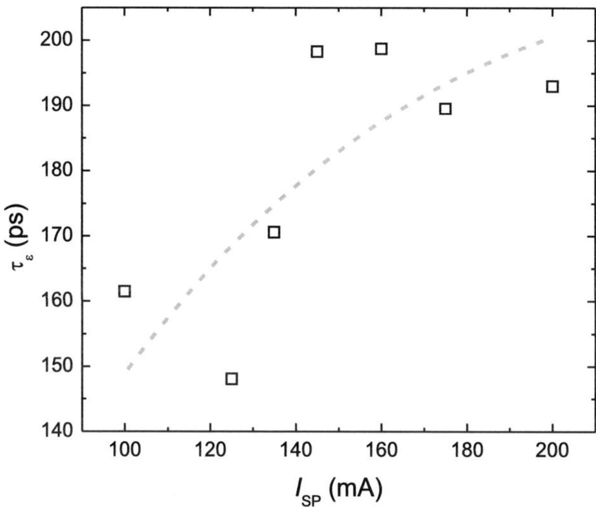

Figure 4.12: Relaxation time constant τ_ε in dependence on the sputter current I_{SP} at which the NbN film with thickness 10 nm were deposited. The dashed line illustrates the general trend.

4.4 Summary

The intrinsic energy realxation time constants of NbN thin films were investigated with measurements in the frequency domain. A measurement setup was assembled, to realize measurements in the optical range and at quasi-equilibrium conditions. The modulated laser light is generated by two slightly detuned laser sources which allow the continuous variation of the modulation frequency in a very wide range. The frequency spectra are analyzed in the framework of the Two-Temperature model: The time-evolution of the photo-response in NbN is dominated by the time constant of the scattering processes between electrons and photons τ_{ep} and the time constant of the phonon escape to the substrate τ_{es}. In NbN thin films, both time constants are closely connected and can not be determined independent of each other. Instead, a combined energy relaxation time τ_ε, which describes the characteristic time of the complete energy relaxation process, is extracted.

In a first series, the dependence of τ_ε on the film thickness was evaluated. The relaxation time constant reduces linearly with the film thickness, which is due to the dependence of τ_{es}. For low film thickness, the mean path for the phonons to the substrate interface is smaller, so the rate of phonon escape is increased. For the smallest film thickness, $d =3.2$ nm, the minimum value $\tau_\varepsilon =124$ ps was achieved. However, for the very thin films the temperature-dependence of the relaxation time constants start to influence on τ_ε, due to the significant reduction of T_C in such films. Additionally, when considering SNSPD operation, the excitation by the absorbed photon is a non-equilibrium situation. In the initial stages, until the energy is distributed inside the electron system, it can thus not be described by the time constants given here. However, this initial redistribution takes place on very short time scales, so that it can be neglected in a first approximation. For a further, more detailed description of the SNSPD pulse relaxation, the description should be expanded to consider the temperature as well as the energy dependence of the involved processes.

In a second series, the influence of the material composition of NbN was investigated. It was found that for increasing niobium content, τ_ε increases. So for minimum SNSPD dead time, the films with reduced Nb content should be preferred. However, τ_ε does not approach the nanosecond range as found for pure Nb films. For all stoichiometric compositions considered here, the values stay in the $100 - 200\,\mathrm{ps}$ range. With these time constants, SNSPD count rates of several GHz can be realized. In typical SNSPD meander configurations, the pulse relaxation is usually limited below this due to the electrical circuitry of device and bias current. However, for applications requiring shorter dead times, this can be easily improved by a change of design to smaller nanowire lengths. To retain reasonable detection areas, this is realized with several parallel SNSPD where one switching nanowire creates an avalanche-like switching of the others [30]. With this approach, the pulse decay time can be reduced down to the level of the energy relaxation time constant. For applications with even stronger demands on the time resolution, a time-correlated single-photon counting (TCSPC) technique can be employed. Here, the resolution is essentially limited by the statistical variance of the response time (jitter) of the SNSPD. However, investigation of those processes requires measurements of time scales below $10\,\mathrm{ps}$, which is not possible with an electrical measurement system as the frequency-domain setup described here.

5 Transport properties of electrons in superconducting nanowires

From the application point of view, a crucial parameter for superconducting devices is the critical current. Almost all of them profit from the ability to bias high currents, and consequently the critical current has been studied intensively since the discovery of superconductivity itself [107]. In the case of nanowire single-photon detectors, higher bias currents lead to higher signal amplitudes which relaxes the requirements on noise prevention in the readout channel. But more importantly, the critical current and its local distribution influence on the detection probability of SNSPD. The detector will only be sensitive in areas where the bias current is close to the local critical current, which can be just a fraction of the detector area if the distribution is strongly inhomogeneous. Furthermore, the spectral bandwidth, which is limited by the minimum required photon energy, depends according to eq. 2.6 on the geometry of the wire (width w and thickness d) as well as the transport current I_B:

$$E_{ph} \propto wd \left(1 - \frac{I_B}{I_C^d} \right).$$

$$(5.1)$$

It is important to note that here, I_C^d is the de-pairing critical current and not the actual critical current of the nanowire, which can be much less than the de-pairing value. Which mechanisms exist that reduce the critical current below I_C^d, and how they can be prevented to retain as large I_B/I_C^d ratios as possible, is the main topic of this chapter.

First, the critical current limiting mechanisms of bridges structured from thin superconducting films are investigated. Models for this special case are presented and then compared with measurements on bridges made from different materials. Here, only straight bridges are considered and w and d are the only variable geometrical factors. Second, a systematic analysis of the influence of the sample layout on the critical current is presented. Nanowires with turns and bends show a current distribution that significantly deviates from that of a straight wire. The quantitative implications on detector properties are discussed and approaches for improvements are formulated.

5.1 Critical current in superconducting bridges

5.1.1 Mechanisms of the limitation of the critical current in superconducting bridges

For the special case of a superconducting bridge made from a thin film, the main scaling parameter defining the properties of the current transport is the magnetic penetration depth λ. Generally, according to the London theory, any magnetic field penetrates into a superconductor over the characteristic length λ_L. The supercurrent flows mainly on the surface of a superconductor with a distribution [108]

$$I(x) = -\frac{H(0)}{\lambda_L} \exp -\frac{x}{\lambda_L}. \tag{5.2}$$

For a 'dirty' superconductor, i.e. where the electron mean free path, l, is much smaller than the coherence length $l \ll \xi_0$, according to Pippard's nonlocal modification of the London model, the magnetic field penetrates over a length $\lambda = \lambda_L(\xi_0/l)^{1/2}$, which ca be much larger than the London penetration depth. Also, the effective coherence length $\xi = (\xi_0 l)^{1/2}$ of such a material is smaller than ξ_0. If one of the dimensions of the superconductor is reduced to or below λ as for the case of thin films, this situation changes. The current distribution normal to the film surface is homogeneous and changes only in the direction along the film with an effective thin film penetration depth $\lambda_{\text{eff}} = 2\lambda^2/d$. It becomes energetically favorable for magnetic vortices to be aligned with their normal core perpendicular to the film surface so that any other orientations can be essentially neglected. If two dimensions are reduced below the penetration depth, as for the case of a nanowire, there is no surface current and the supercurrent will be transported by the wire with a homogeneous distribution across the wire cross-section.

For the generation of the critical state in a current-carrying wire, there are two principal mechanisms: The first is the unbinding of the Cooper pairs that carry the super-current. This happens when the current density becomes so high that the required Cooper pair velocity yields a kinetic energy that exceeds the binding energy. This is called the de-pairing critical current density j_C^d and marks the ultimate limit of any superconducting current transport. The de-pairing critical current density j_C^d can be calculated from the material properties according to Ginzburg-Landau:

$$\begin{aligned}
j_C^d(T) &= j_C^d(0) \left(1 - \frac{T}{T_C}\right)^{\frac{3}{2}} \\
j_C^d(0) &= \frac{4\pi(e^\gamma)^{\frac{3}{2}} (\Delta(0))^{\frac{3}{2}}}{21\varsigma(3)\sqrt{6}\, e\rho \sqrt{D\hbar}}.
\end{aligned} \tag{5.3}$$

Here, $\varsigma(3) = 1.202$ and $\gamma = 0.577$; $\Delta(0)$ is the energy gap at zero temperature, ρ is the normal state resistivity, e is the electron charge, $\hbar$ is Planck's constant and D is the electron diffusivity. In the vicinity of T_C the temperature dependence $(1 - T/T_C)^{3/2}$ has been shown to describe experimental data very well. For the full temperature range, a correction factor is applied that follows from a rigid numerical calculation for the dirty limit [109].

In type-II superconductors, especially for thin films, the second mechanism is the penetration and movement of magnetic vortices into the bridge. In this case, there is a competition between the Lorentz force exerted from the applied current on the vortex and the pinning forces given e.g. by material imperfections. According to Likharev [110], magnetic vortices can not penetrate into superconducting bridges with a width w smaller than $4.4\xi_{GL}(T)$, where $\xi_{GL}(T)$ is the temperature dependent Ginzburg-Landau coherence length. So in the case of a nanowire with a width so small that $w < 4.4\xi_{GL}(T)$ in the whole temperature-range, the de-pairing critical current density j_C^d can be observed. But even for wider bridges, at temperatures approaching T_C, the coherence length diverges and the vortices are expelled. Thus, the critical current density is also limited by the de-pairing critical current in the vicinity of T_C. Contrarily, for low temperatures the condition $w < 4.4\xi_{GL}(T)$ is not satisfied and penetration, de-pinning and movement of magnetic vortices can significantly reduce j_C.

Ideally, the condition for vortices to penetrate into the bridge is the presence of a magnetic field that exceeds the first critical magnetic field of the superconductor, which can be estimated [108] as

$$B_{C1}\left(\frac{T}{T_C}\right) = \frac{\Phi_0}{4\pi\lambda^2}\left(\ln\frac{\lambda\left(\frac{T}{T_C}\right)}{\xi_{GL}\left(\frac{T}{T_C}\right)} + 0.08\right). \tag{5.4}$$

Here, Φ_0 is the magnetic flux quantum and λ the temperature dependent magnetic penetration depth. Even without any external magnetic field, there is still the magnetic field caused by the applied transport current density j_{tr} leading to so-called self-generated vortices. The field created by this current at the edges of a superconducting bridge is

$$B_{\text{edge}} = \frac{\mu_0 j_{tr}\sqrt{wd}}{2\pi}, \tag{5.5}$$

where μ_0 is the vacuum permeability [111]. In a non-ideal bridge, there is a probability for vortices to enter even for magnetic fields much lower than B_{C1}. The intrusion of vortices into a thin film can be described by an effective energy barrier in a very similar manner as for bulk materials [60]. In the case of small samples with dimensions below λ_{eff}, this barrier becomes one-directional - it prevents vortices from entering the bridge but allows their exit [112]. The vortex has a certain probability to overcome this barrier and enter the bridge. This probability is strongly dependent on the uniformity of the bridge edges. Defects caused by the etching process or non-uniformities of the material can lead to local differences in the edge shape or structure which reduce the energy barrier.

Once the vortices penetrate into the wire, the critical state generation depends on their dynamics. The current flowing through the wire exerts a Lorentz force $\mathbf{F}_L = \mathbf{j} \times \mathbf{B}$ per unit volume in the direction along the bridge width. In the case of sufficiently strong pinning, they are static

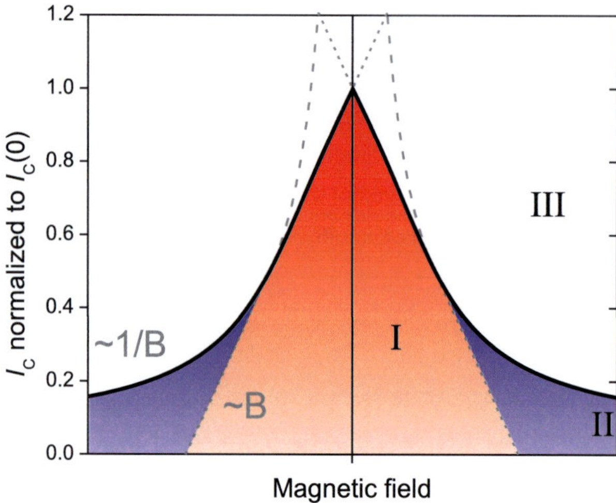

Figure 5.1: Schematic dependence of normalized critical current $I_C/I_C(0)$ of a superconducting bridge on an external magnetic field B. For small fields, the bridge is initially in the Meissner state (zone I) and the critical current that marks the transition to the resistive state (zone III) decreases linearly with increasing magnetic field. For higher magnetic fields, the bridge is without field in the static mixed state (zone II) and the transition follows a $1/B$-dependence

and only reduce the effective area for the current flow by their presence. The size of the normal core of a vortex is roughly $2\xi_{GL}(T)$, so depending on the density of vortices and the temperature, the measured critical current density is appropriately lower than what would be expected from the geometrical dimensions. However, the critical state is still determined by the de-pairing of Cooper pairs once j_C^d with respect to the reduced area is reached. If the pinning force is exceeded by the Lorentz force acting on the vortices in the presence of a transport current, the vortices will start to move and induce an electric field $\mathbf{E} = \mathbf{v} \times \mathbf{B}$, where $\mathbf{v}$ is the velocity of the vortices. The movement of the normal cores results in an energy dissipation that causes the critical state. The critical current density is the current density for which the vortices start to move (i.e. for which a detectable voltage is produced across the superconductor) and is called the de-pinning current density j_C^p. For a material with weak pinning, j_C^p can be much smaller than j_C^d.

Critical current of superconducting nanowires in magnetic field

In addition to the self-generated magnetic field, the critical current is further reduced if an external magnetic field is applied. In the absence of bulk pinning, the case is well described theoretically [113] for relatively wide nanobridges ($w \gg \lambda_{\text{eff}}$) in a field perpendicular to the sample surface.

A typical dependence of the critical current on the magnetic field strength $I_C(B)$ is shown in Fig. 5.1. For small fields, the film is initially in the Meissner state. Once the combined external magnetic field and the self-field of the current are large enough to allow vortices to penetrate into the superconductor, the vortices immediately start to move and the bridge switches to the normal state. In this regime, $I_C(B)$ decreases linearly with increasing magnetic field. The slope of this reduction depends on the width of the bridge, and for the wide bridge limit the derivative is directly proportional to the width: $dI_C/dB = -w/(4\mu)$. For larger applied magnetic field, the vortices may enter already at low currents, where (for small pinning) they are driven by Meissner currents into the center of the bridge and form the static mixed state. This state is stable until the current becomes large enough for the vortices to move and the film enters the dynamic mixed state. This current defines the critical current for the high magnetic field range and follows a dependence $I_C(B) \propto 1/B$. The magnetic field where the behavior crosses over from the linear to the $1/B$-dependence is to a good approximation at $I_C(B) = 0.5 I_C(0)$ if $\xi \ll w$.

For smaller sample widths ($w \approx \lambda_{\text{eff}}$) the description was extended to include the already significantly altered current distribution across the bridge [114]. In this case, the derivative dI_C/dB for low fields shows a deviation from the linear width dependence which becomes more apparent the smaller the sample width [114]. The predictions were experimentally verified on bridges made from 200 nm thick MoGe, a material with low intrinsic pinning.

If the width of the sample is further reduced to the range $w \ll \lambda_{\text{eff}}$ we finally arrive at the situation where the current is distributed homogeneously across the cross-section of the wire. This limit was described in [115], where for a bridge initially in the Meissner state, the derivative of the linear dependence close to $B = 0$ was predicted to be

$$\frac{dI_C}{dB} = -\frac{cw^2}{4\pi\mu\lambda_{\text{eff}}}. \tag{5.6}$$

In summary this means that for narrow bridges, the magnetic field dependence becomes weaker and the structures are less susceptible to external field. In the case of nanowires of a sufficiently weak pinning material, the expected dependence of the critical current is for small fields linear. The slope of this dependence should rise quadratically with the width of the nanowire.

5.1.2 Critical current in Nb bridges

To experimentally investigate the theoretical dependences discussed above, niobium films are considered. The properties of niobium are well studied, enabling quantitative comparison with theory and literature values. The material can be deposited into thin films that show especially high uniformity when the substrate is heated during deposition, i.e the electron mean free path l is much larger than in NbN [116]. The material behaves more closely to an ideal type-II superconductor and is a good starting point for a systematic study of transport currents.

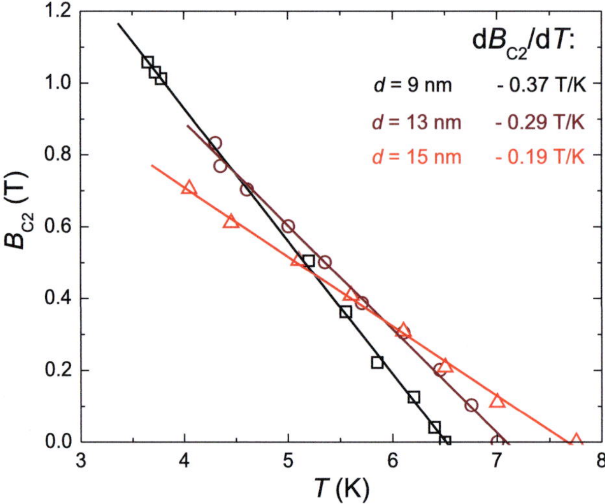

Figure 5.2: Temperature dependence of the second critical magnetic field B_{C2} measured on the unstructured Nb films with different thickness d. The solid lines show linear fits to the data from which the slope dB_{C2}/dT is extracted.

The Nb films were deposited by magnetron sputtering very similar to the description of section 3.1, without the reactive gas component. The sapphire substrate was in this case heated to 800 °C during deposition. The results discussed in this chapter were published in [77], where more details on the Nb film deposition and parameters can be found.

To study the critical current, three films with thickness d =9, 13 and 15 nm were deposited. Their critical temperatures were evaluated to be T_C =6.1, 7.2 and 8.2 K, respectively. Before patterning, the temperature-dependence of the second critical magnetic field B_{C2} was evaluated (see Fig. 5.2) which shows a linear dependence in the full range of measurement typical for Nb. By the same methods described in section 3.3, the electron diffusion coefficient D is evaluated. The coherence length is estimated to $\xi_{GL}(0) = 14 - 17$ nm, depending on the film thickness. Each film is then structured by electron-beam lithography into bridges with different widths between 100 nm and 10 µm. Comparing those with $\xi_{GL}(0)$, we see that for all bridges the first criterium for vortex entry described above is principally fulfilled. We thus expect to measure a j_C which is either a) the pair-breaking current density or b) the penetration current density, if the induced magnetic field is sufficient for the vortices to overcome the edge barrier, or c) the depinning current density if condition b) is met but strong pinning forces are present.

The critical current densities measured at 4.2 K are shown in Fig. 5.3. Thicker films generally have a higher critical current density due to the proximity effect, as described for the case of NbN in section 3.1.2. The same effect also reduces $j_C(4.2\,\mathrm{K})$ for very narrow bridges and leads

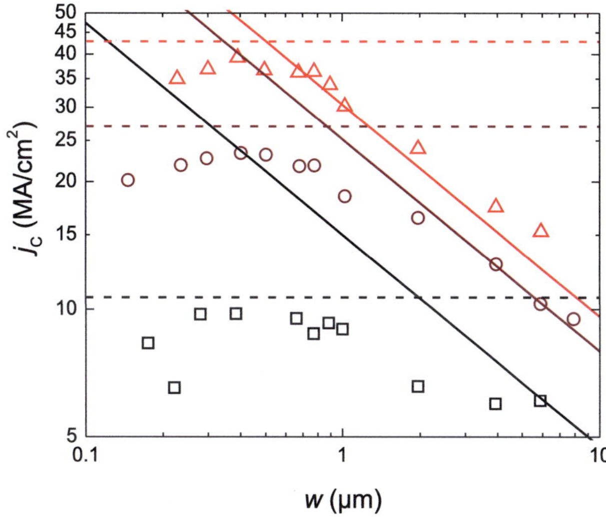

Figure 5.3: Critical current densities j_C measured at 4.2 K in dependence on the width of the Nb bridges. The squares are data from the 9 nm thick film, circles from 13 nm and triangles from 15 nm. The dashed lines mark the pair-breaking critical current limit calculated for each film from eq. 5.3. The solid lines represent the width dependence of half the penetration current density $0.5 j_{pen}$ according to eq. 5.7.

to the slight reduction of the critical current density for $w < 400$ nm observed in all films. Apart from this observation, $j_C(4.2\,\mathrm{K})$ is almost constant for bridges with $w < 1$ μm. For larger w, the dependence steadily decreases for all films and seems to be steeper for larger film thickness. This cross-over at $w = 1$ μm marks the change of the limiting mechanism for the critical current in the bridges.

To investigate this more closely, Fig. 5.4 shows the full temperature dependence of j_C of a few selected samples with film thickness $d = 13$ nm. The dependences look very similar for other film thicknesses, only scaled for the different absolute values of T_C and j_C. Shown are two samples from the 'constant' region with widths 200 and 400 nm, and three samples from the micrometer region with 2, 4 and 10 μm. The j_C values at very low temperatures reflect the behavior shown in Fig. 5.3. However, for temperatures close to T_C, all curves follow the same dependence.

From the measured parameters T_C and ρ and the electron diffusion coefficient D, which was extracted from the magnetic field measurement, the de-pairing critical current can be calculated according to eq. 5.3. The respective temperature dependence is shown by the dashed line in Fig. 5.4. It fits remarkably well to the measurement values taken in the vicinity of T_C for all samples, affirming that here the critical current is indeed limited only by the pair-breaking current. For lower temperatures, the data diverge from the theoretical description with a strong dependence on the width. While the narrow samples follow j_C^d in a wide temperature range

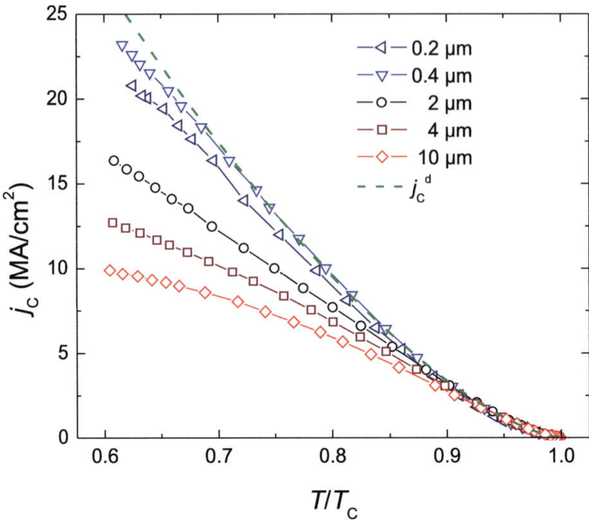

Figure 5.4: Temperature dependence of the critical current density $j_C(T)$ of samples with different width from the 13 nm thick Nb film. The dashed green line shows the de-pairing critical current density calculated according to eq. 5.3.

down to $0.7T_C$, the wide bridges deviate already at higher temperatures and stay farther below the de-pairing current. For the largest bridge there is even a saturation of j_C visible in the low-temperature region.

In this region, the critical current is reduced by the penetration of magnetic vortices into the film. The only source of magnetic field in this measurement is the edge field generated by the applied current itself. According to eq. 5.5, $B_{\text{edge}} \propto \sqrt{w}$. Unfortunately, the exact value of the penetration field required to overcome the vortex entry barrier is unknown. As a rough estimation, we assume that it is equal to the first critical magnetic field B_{C1}, as would be the case for a very wide bridge. Since B_{C1} is a material property, it is itself independent of the bridge width. From eq. 5.5 and eq. 5.4, we then arrive at an estimate for the penetration current

$$ j_{pen} = \frac{2\pi B_{C1}}{\mu_0 \sqrt{wd}}. \tag{5.7} $$

The values of j_{pen} calculated for the Nb bridges at 4.2 K are by roughly a factor of 2 larger than the measured j_C. However, the dependence on width (solid lines in Fig. 5.3) fits quite well and explains the reduction of j_C for the wider bridges: Due to the larger cross section, they can support higher absolute transport currents that generate a larger magnetic field. This field causes vortices to penetrate into the bridge, which are immediately set into motion by the Lorentz force. The de-pinning current density seems to be lower than j_{pen}, which is reasonable for Nb films

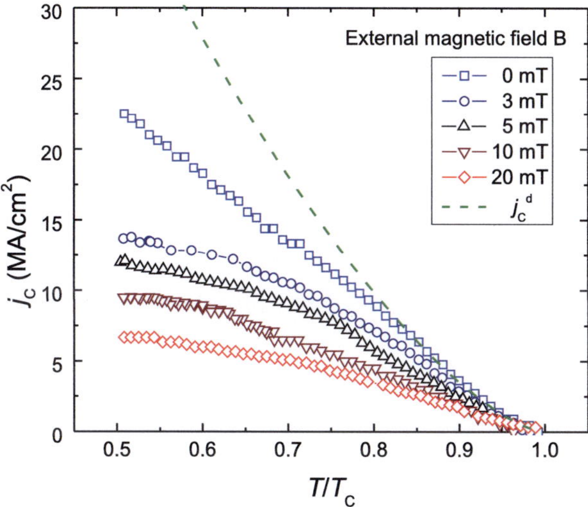

Figure 5.5: Temperature dependence of the critical current density $j_C(T)$ of a Nb bridge with $d = 15\,\text{nm}$ and $w = 2$ µm in an external magnetic field applied perpendicular to the sample surface. The dashed line reflects the theoretically calculated temperature dependence of the de-pairing critical current density without magnetic field.

with high uniformity. For the bridges with $w < 1$ µm, the self-generated magnetic field becomes so small that the de-pairing critical current density is reached before j_{pen}.

The assumption that j_C is limited by the penetration field is further confirmed by the measurement of $j_C(T)$ in an external magnetic field. Fig. 5.5 shows such a measurement for the sample with large cross-section $d = 15\,\text{nm}$ and $w = 2$ µm. The field is applied normal to the substrate surface. We note that here the curves do not run as close together for temperatures near T_C as was the case without external field. This can be partly explained by the reduction of T_C by the magnetic field and consequently also the reduction of j_C^d. In Fig. 5.5, only the j_C^d-dependence for the case without external field is shown as a dashed line. The curves show otherwise a qualitatively very similar suppression of j_C with increasing magnetic field as the curves in Fig. 5.4 with increasing bridge width. Remarkably, this reduction is already significant for magnetic fields $B < 3\,\text{mT}$, which is smaller than B_{C1} used for the estimation above. Thus, we can assume that j_{pen} was overestimated and that the real value is closer to the measurements. Unfortunately, 3 mT is the minimum resolution of the magnetic field in the experimental setup, so it is not possible to determine the penetration field directly.

In conclusion, we find that the limitation of the critical current density in Nb bridges can be well understood by the considerations given in section 5.1.1. Close to the critical temperature, the current can always reach the de-pairing critical current. The temperature region for which this is given is increased as the width of the sample becomes smaller. For very narrow samples with a few 100 nm width even at 4.2 K the de-pairing critical current marks the limit of j_C. For wider bridges, the self-generated magnetic field caused by transport currents above j_{pen} allows vortices to penetrate into the film where they are accelerated by the Lorentz force and the superconductivity breaks down before j_C^d is reached. It has been shown [117] on bridges of similar width made from slightly thicker ($d = 20$ nm) Nb films, that the reduction of j_C by vortex entry can be avoided if the critical current is measured by a pulsed current method. The de-pairing critical current can then be reached also for samples several μm wide. However, for most nanowire device applications, such an operation is not feasible. Thus, to achieve critical current densities for constant applied currents close to the de-pairing limit, superconducting wires should be in the sub-μm range in width and have a high edge barrier for vortex penetration.

5.1.3 Critical current in Ba(Fe,Co)$_2$As$_2$ bridges

A material that behaves rather differently than Nb concerning current transport is Ba(Fe,Co)$_2$As$_2$ (Ba-122), one of the recently discovered ferroarsenide-based superconductors. Recent advances in thin film and structuring technologies of such materials put them into perspective for possible future application in photon detectors. Since these materials have a very different layered structure than the common low-temperature superconductors, it could be expected that the current transport in structured samples provides interesting properties. In the following, the material will be briefly introduced and an overview of relevant preliminary work is given. Then, the experimental investigation of the current transport properties is described and the results discussed in view of application-relevant properties.

Iron-based superconductors

A new class of high-T_C superconducting materials was recently discovered and named iron-based superconductors or ferropnictides [118]. Their distinct feature is a layer of iron and arsenic (or other pnicogen or chalcogen) tetrahedrons in its chemical compound which is responsible for the main electronic transport. Like in the cuprate superconductors, these layers alternate with the other, more isolating layers of the material, creating an intrinsic anisotropy in the current transport. First superconductivity was found in 2008 in LaOFeAs, where a critical temperature of 26 K was reported [118]. This paved the way for the discovery of numerous similar compounds, where T_C values up to 55 K were reported. In many such materials, the superconducting state can be improved or only established by doping of the compound. The materials are roughly classified by their chemical stoichiometric formula indices. The most common families are:

- The '1111'-family or oxypnictides, first reported compound La$_1$O$_1$Fe$_1$As$_1$

- The '122'-family, first reported compound Ba$_1$Fe$_2$As$_2$ [119]

- The '111'-family, first reported compound $Li_1Fe_1As_1$ [120]

- The '11'-family, first reported compound Fe_1Se_1 [121]

These materials provide a new set of challenges for the field of superconducting thin film growth. Successful highly-epitaxial thin film ($d \leq 100\,nm$) deposition has been demonstrated for the $Ba_1Fe_2As_2$ compound [122], in the following abbreviated as Ba-122. Bulk samples with hole-doping ((Ba_{1-x},K_x)Fe_2As_2) show superconducting transition temperatures of up to $T_C \approx 38\,K$ [119], with electron-doping ($Ba(Fe_{1-x},Co_x)_2As_2$) up to $22\,K$ [123]. This group of ferropnictides has been suggested as most suited for application purposes for various reasons [124]: They have rather high critical temperature and upper critical field $B_{C2} \geq 45\,T$ [125], low electronic anisotropy, and an apparently strong pinning of magnetic vortices. Also it is relatively well to handle, for example it was shown that Ba-122 is more stable against exposure to moisture than Sr-122 [126], and the mechanical and chemical stability allows the use of lithography techniques for patterning. Finally, the preparations of thin films made from the 122-family are easier than from the F-doped LnFeAsO compounds (Ln being a rare earth element). For this reason, the Co doped Ba-122 was selected for the following investigation. It has been shown recently, that the superconducting properties of the Ba-122 films can be further increased by an iron buffer layer between film and substrate due to an improved lattice matching [127]. As the additional iron layer would complicate the analysis, it is not considered here.

The critical current of Ba-122 was previously experimentally investigated by a variety of different techniques [128]. j_C in $Ba(Fe,Co)_2As_2$ evaluated by magnetization measurements in single crystals are found to be $3 - 4 \cdot 10^5\,A/cm^2$ at $4.2\,K$ [129, 130]. Early results obtained on thin films were typically below $10^5\,A/cm^2$ [131]. However, a j_C value of $4 \cdot 10^6\,A/cm^2$ has been recently reported for a $250\,nm$ thick epitaxial $Ba(Fe,Co)_2As_2$ film on $(La,Sr)(Al,Ta)O_3$ (LSAT) (001) substrate [132] and similar values are found in $Fe/Ba(Fe,Co)_2As_2$ bilayers [131]. Comparable j_C values were reported for composite films including correlated defects which may enhance pinning [133, 134].

These measurements were achieved on either bulk or thin film samples, where no influence of a finite sample width on the critical current density is present. To study the applicability of this material class for real devices, it is imperative to investigate the transport properties directly on structured samples. In this section, the critical current density dependence of structured Ba-122 bridges will be presented and discussed. This work on μm-sized Ba-122 bridges was published in [135] and was later extended into the sub-μm range in [136]. To the best of our knowledge, it was the first report of transport current measurements on structured Ba-122 bridges. First reports of a Josephson Junction made from Ba-122 and correspondent measurements were shown by [137] and first photo-response measurements of detector structures are reported in [138].

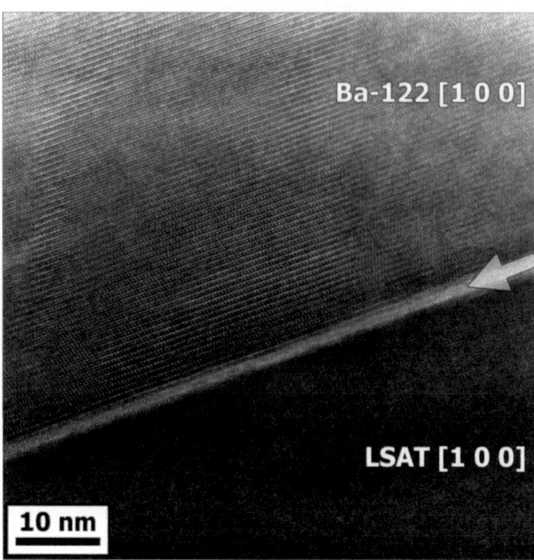

Figure 5.6: TEM image showing the highly epitaxial Ba-122 film (top) on the LSAT substrate (bottom). At the interface to the LSAT substrate, a thin iron rich interfacial layer is formed indicated by the arrow [135].

Ba(Fe,Co)$_2$As$_2$ thin films

The transport current is experimentally investigated on Co-doped Ba-122 films. The films were grown and characterized at the Institute for Metallic Materials of the IFW in Dresden. They are deposited by pulsed-laser deposition under ultra-high vacuum condition with a base pressure of 10^{-9} mbar in the process chamber. The nominal target composition Ba(Fe$_{0.9}$Co$_{0.1}$)$_2$As$_2$ is transferred from the target to the film during the PLD process, which was confirmed by Rutherford Backscattering Spectrometry. The film was confirmed to be c-axis textured by XRD in Bragg-Brentano geometry with Co-Kα radiation. More details on the thin film preparation can be found in [122]. The samples were grown with thicknesses between 50 and 100 nm on LSAT (100) substrates that were kept at 650°C during deposition. A lamella of the film/ substrate cross-section was prepared on a Carl Zeiss 1540 XB focused ion beam using the in-situ lift-out method. The microstructure of the film could then be investigated with a C$_s$-corrected FEI Titan^3 transmission electron microscope (TEM) operating at 300 kV (see Fig. 5.6). The TEM investigation confirms the epitaxial growth of the Ba-122 films and is used to determine the exact film thickness d.

Before structuring, the critical temperature of the films were evaluated by a standard four probe measurement technique in a commercial Physical Properties Measurement System in a temperature range around the superconducting transition. Additionally, a series of $R(T)$ curves under the influence of magnetic field was measured in this setup on a film with a T_C of 20.2 K (see Fig. 5.7 (left)). The external magnetic field up to 9 T was applied normal to the sample surface and causes a broadening of the resistive transition with increasing applied magnetic field. The transition width $\Delta T = T(0.9R_{30}) - T(0.1R_{30})$, where R_{30} is the normal state resistance at

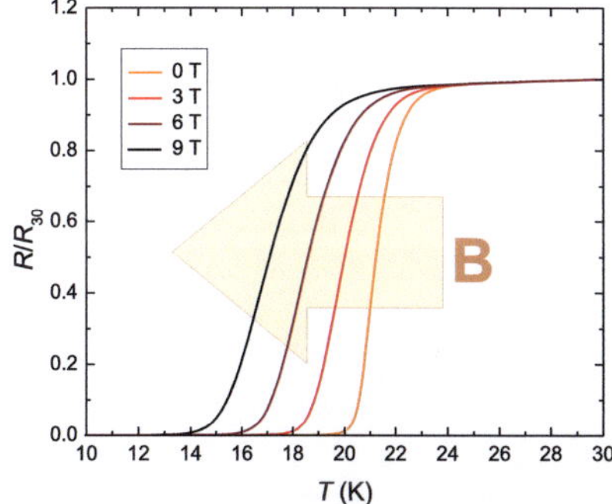
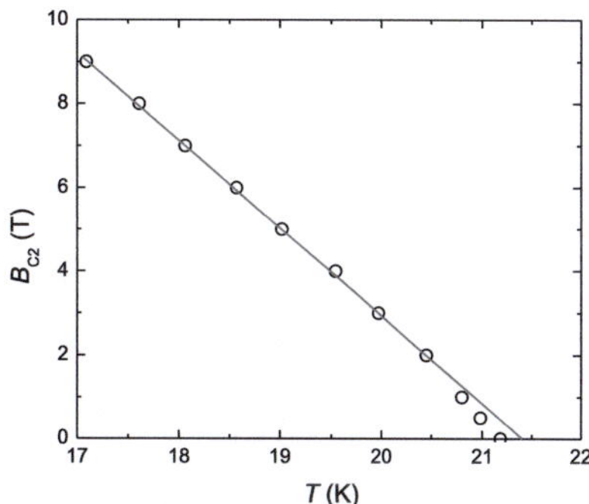

Figure 5.7: Left: Normalized resistance over temperature for an unstructured Ba-122 thin film in external magnetic field B. The superconducting transition becomes broader the higher the applied field. For each temperature, the magnetic field for which the resistance reached $0.5R_{30}$ was evaluated as the second critical magnetic field. The resulting temperature dependence is shown the right graph. The gray line is a linear fit to the data at low temperatures.

$T = 30\,\mathrm{K}$, increases from about $1.8\,\mathrm{K}$ at zero magnetic field to almost $4\,\mathrm{K}$ at $9\,\mathrm{T}$. To determine the upper critical field B_{C2} without a too strong influence of the transition broadening, it was defined as the magnetic field for which the sample resistance was half of the normal state resistance $B_{C2} = B(R = 0.5R_{30})$. The temperature dependence $B_{C2}(T)$ is shown in Fig. 5.7 (right). At low temperatures (approximately below $0.9T_C$), the dependence is linear (solid line) with $\mathrm{d}B_{C2}/\mathrm{d}T = -2.06\,\mathrm{T/K}$, which is similar to the value -2.5 T/K reported for the bulk material [129]. The strong broadening of the transition and the deviation of B_{C2} from the linearity close to T_C are indications for a granular structure of the film. With eq. 3.4 and eq. 3.5, we can estimate $B_{c2}(0) = 31\,\mathrm{T}$ and the corresponding coherence length $\xi_{GL}(0) = 3.3\,\mathrm{nm}$ from this dependence.

Development of the patterning process

To determine the transport current properties of the Ba-122 thin film, a lithography process has to be developed to structure the films into bridges with a well-defined geometry. Several unique requirements on the process arise from the material which have to be fulfilled by a modification of the standard lithography process

- The Ba-122 compound is vulnerable to many chemical substances used in lithography processes, including water and acetone

- To keep its stoichiometric composition, the film must not be exposed to high temperatures. An arbitrary limit of 100°C was set for the processes and proved to be adequate.

- The stability of the thin films does not allow any kind of mechanical cleaning treatment, including supersonic bath.

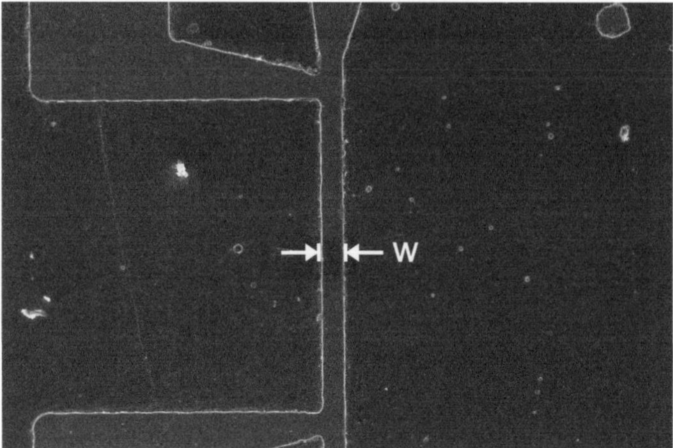

Figure 5.8: SEM image of one micro-bridge structured from a Ba-122 thin film. The image shows the bridge and the four contacts for the current (top and bottom) and the voltage (left). The whole surface was covered by a thin ($\approx 10\,$nm) film of Nb to prevent charging effects during the electron-beam microscopy.

- The film should not be exposed to ambient atmosphere for extended periods of time.

Due to these restrictions, it was not possible to reasonably clean the sample surface prior to the lithography process. This results in a high number of dirt particles in the range from 200 nm to several µm in diameter, which significantly reduced the yield of unaffected samples.

In the case of bridge widths 2 to 10 µm, a one-step photolithography process can be employed. The AZ5214 resist allowed out-baking at relatively low temperatures of 85°C to prevent detrimental effects on the film. The exposure to a UV light source was made through a chrome mask in contact with the resist surface. Although the AZ developer is water based, which is known to have a negative influence on Ba-122, the remaining resist protects the main part of the bridges. Only at the sides of the bridges a damaging influence of the water is expected. Also, in the following ion-beam etching process, the resist with a thickness of ≈ 1.2 µm protected the film from physical damage or ion implantation. The etching process was performed by Ar ions accelerated in an electric field with $U = 250$ V with a power of $P = 185\,$W, during which the samples is water-cooled to ≈ 10°C to prevent over-heating. Once again, the sides of the samples remain unprotected and subsequent SEM imaging confirmed a relatively strong edge roughness. The SEM image of a Ba-122 bridge structure with $w = 2$, µm is shown in Fig. 5.8. For samples where the width is much larger than the film thickness, the influence of the edges results only in a minor decrease of T_C. However, for the development of a patterning process for much smaller sample widths as would be required for many applications, this issue needs to be addressed further. An improved structuring process with electron-beam lithography allowed to reduce the bridge width below 1 µm. Here, a negative exposure process was used because it allowed the bake-out

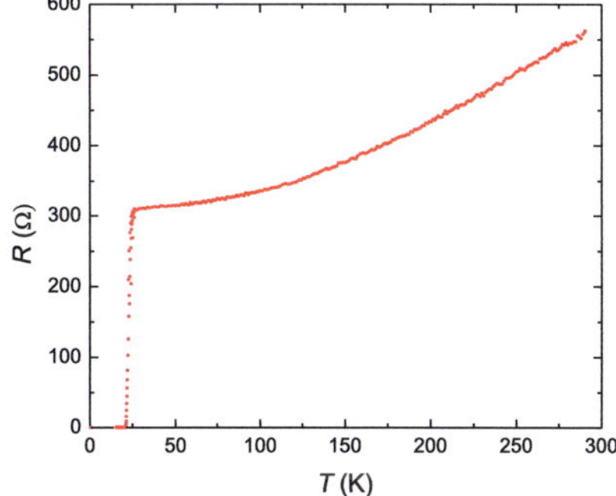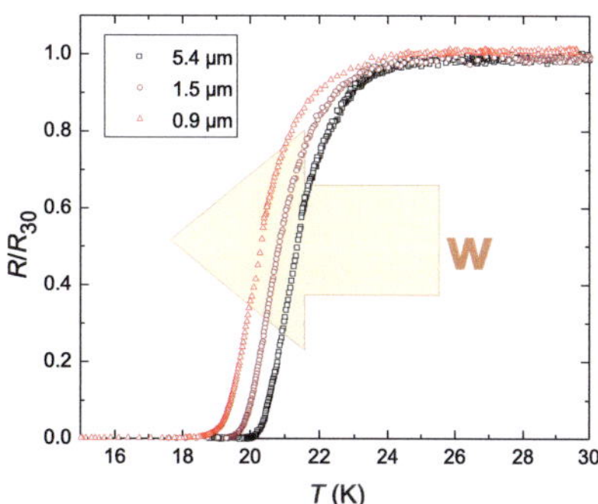

Figure 5.9: Left: Typical dependence of the sample resistance on temperature in the full range from room temperature to 4.2 K. In this case, $w = 2.9\,\mu$m. Right: Resistance over temperature around the superconducting transition of three Ba-122 bridges with different width structured from a 50 nm thick Ba-122 film.

of the resist AR-N 7520.18 at 85°C as in the case of the photolithography process. After the electron-beam exposure the resist was developed in AR 300-47 and stopped with distilled water. Once again the contact with water is unavoidable and is kept as short as possible. The sample is then etched by the same ion-beam etching process as described above. A detailed description of the structuring processes developed for the Ba-122 thin films and additional processes for the structuring of bolometer detectors from these films can be found in [139].

After structuring, the samples were wire bonded in a four-probe configuration and the temperature dependence of the resistance was measured from room temperature to 4.2 K in a liquid Helium Dewar. Figure 5.9 (left) shows one typical representative of $R(T)$ in the full temperature range. The resistance decreases with the temperature like in a metal, until it reaches the superconducting transition temperature. The reduction of the resistance from 295 to 30 K was found to be $R_{295}/R_{30} \approx 1.9$, which is similar to what has been reported for unstructured films [122]. Fig. 5.9 (right) shows a close up of the transition for three selected bridges with the width of 0.9, 1.5 and 5.4 µm structured from a 50 nm thin Ba-122 film. The critical temperature of the widest samples $T_C = 19.8$ K was reduced only about 0.5 K as compared to the unstructured film, which shows that the structuring process has no major detrimental effect on the superconducting properties of the Ba-122 film. Also, the transition width $\Delta T = 2.2$ K, is almost the same ΔT value as for the non-patterned film. With decreasing w the transition temperature is shifted to lower temperatures, without a change in the transition width. The reduction of T_C values for narrow structures is similar to what has been found for Nb bridges. From the resistance of the bridges measured at $T = 25$ K and their geometry, the residual resistivity was calculated to $\rho_{25} = 301\,\mu\Omega$cm, which is comparable to what was reported previously [125, 132, 133]. Multiple measurements of the microbridges were made within a few months after patterning. Although the microbridges were thermo-cycled several times, both T_C and j_C at 4.2 K always showed the same values within

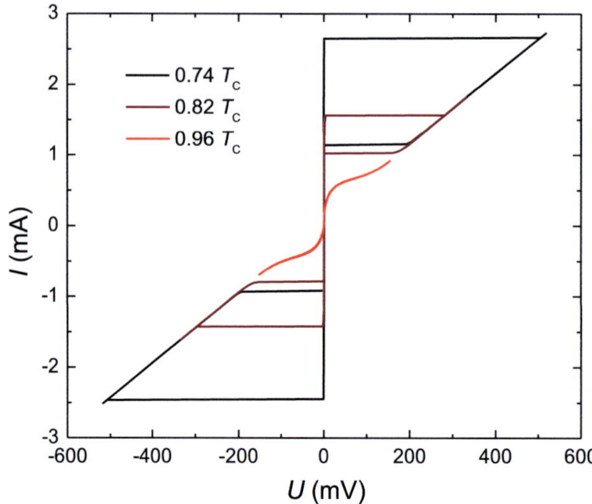

Figure 5.10: Current-voltage characteristics for three selected temperatures measured on a Ba-122 bridge with $w = 4.7$ µm, structured from a 90 nm thick film. The curves show a clear jump of the sample to the resistive state once the critical current is reached. For temperatures around T_C, the transition becomes smooth and gradually changes from an S-shape to the ohmic linear dependence.

the measurement accuracy, demonstrating good long-term as well as thermal stability. With an energy gap value $\Delta(0) \approx 1.1 k_B T_C$ [140], the London penetration depth can be estimated with $\lambda^2(0) = \hbar \rho_{30}/(\mu_0 \pi \Delta(0))$. For the 50 nm thick film we arrive at an effective thin film penetration depth of $\lambda_{\mathrm{eff}} = 2.31$ µm. This means that in this case, the bridges have widths $w \approx \lambda_{\mathrm{eff}}$ and we can expect some non-uniform current distribution across the wire cross-section.

Critical current of bridge structures

In the same experimental setup the current-voltage characteristics were measured at temperatures from 4.2 K to T_C to determine the critical current (see Fig. 5.10). For low temperatures, the switch into the resistive state occurs as a sharp jump at the critical current. As the temperature approaches T_C, the hysteresis becomes smaller and the transition smoother. For $T > 0.85 T_C$, the critical current is defined as the current for which the voltage rises above 100 µV, which is the voltage level slightly above the resolution of the measurement system. At $T = 4.2$ K, j_C values of up to 3.2 MA/cm² are achieved, which is significantly higher than the value reported for single crystals [129, 130] and comparable to recently reported values of 4 MA/cm² measured in a 250 nm thick Co-doped Ba(Fe,Co)$_2$As$_2$ epitaxial film [132] or obtained in Fe/Ba(Fe,Co)$_2$As$_2$ bilayers [131].

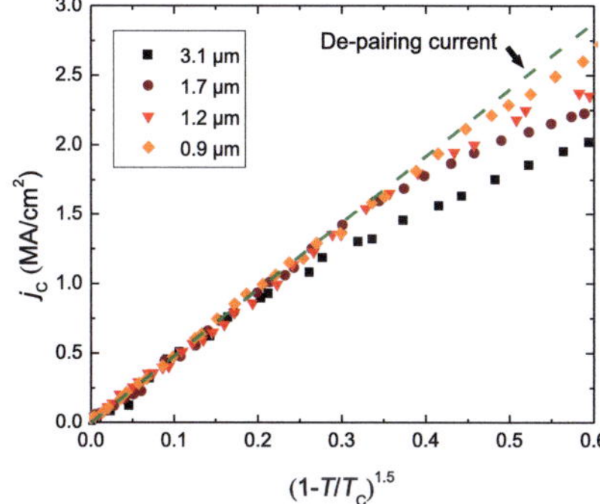

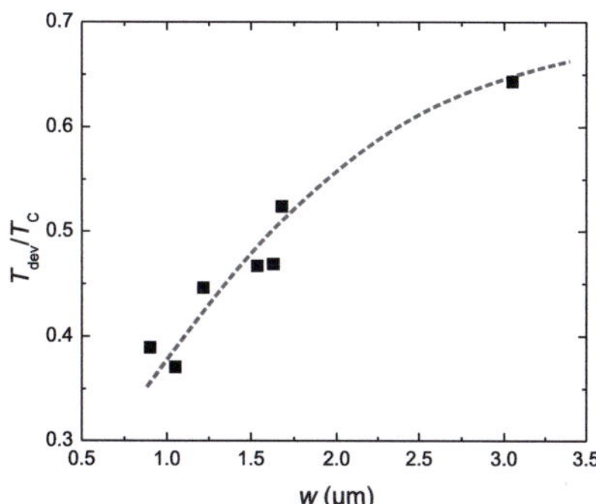

Figure 5.11: Left: Critical current over the temperature scale $(1 - T/T_C)^{3/2}$ for several bridges of different widths structured from a 50 nm thick Ba-122 film. The dashed green line represents the dependence of the de-pairing current. The temperature for which the data of a bridge deviate from the line is called T_{dev}. Right: Dependence of T_{dev} relative to T_C for all measured bridges. The trend is illustrated by the dashed line.

To compare the temperature dependence of j_C with the de-pairing critical current dependence, the values are plotted in the temperature scale of j_C^d according to eq. 5.3. The de-pairing limit is then linear, as shown by the solid line in Fig. 5.11 (left). For temperatures $T \approx T_C$ (left side of the graph), the j_C values follow this dependence well. At low temperatures (right side), the measured j_C eventually deviate from the linear dependence at a temperature T_{dev}. For samples with smaller w, the data follow the de-pairing behavior over a larger temperature range and T_{dev} is lower, as was observed for the Nb bridges. The values of T_{dev} relative to T_C for all measured samples are shown in Fig. 5.11 (right). Although the principle behavior is the same as for the Nb bridges, the range of the temperature in which the de-pairing current is reached is significantly wider, especially for µm-sized bridges. In Nb, the smallest value for $T_{dev}/T_C \approx 0.7$ was achieved only for 200 nm wide bridges, whereas in pnictides, this value is already reached for the 3 µm bridge. For the smallest bridge with $w = 900$ nm, T_{dev}/T_C values below 0.4 can be realized.

The coherence length of the Ba-122 ($\xi_{GL}(0) = 3.3$ nm) and also $4.4\xi_{GL}(0)$ is much smaller than the sample width. Even for the smallest bridge, Likharev's criterion would limit the entry of vortices only for the very narrow temperature interval $T > 0.997 T_C$, so there must be another mechanism preventing vortices from limiting the critical current. In the case of the Nb bridges, similar conditions were observed and the deviation of j_C from the de-pairing limit at low temperatures was attributed to overcoming of the vortex entry barrier by the self-generated magnetic field of the bias current. As a first rough estimation of the edge barrier of the Ba-122 film, we once again estimate the first critical magnetic field according to eq. 5.4 to be $B_{C1}(0) = 33$ mT, which is in the same order of magnitude as $B_{C1}(0) = 22$ mT of Nb. However, the critical current density of about 3 MA/cm^2 at 4.2 K is much lower than what is measured on a Nb bridge of similar size. Thus, even the Ba-122 bridges with comparatively large widths have a self-field

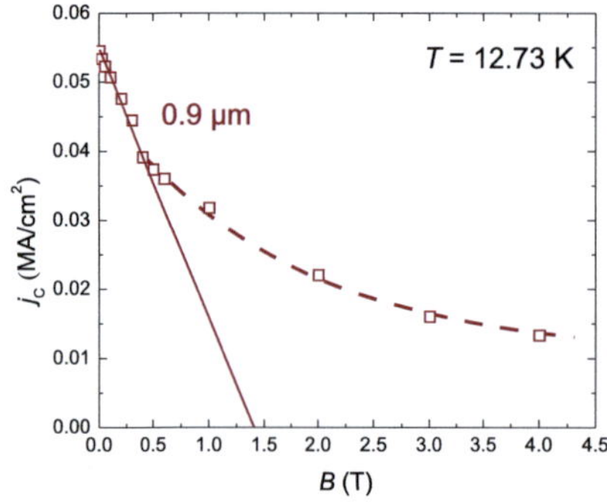
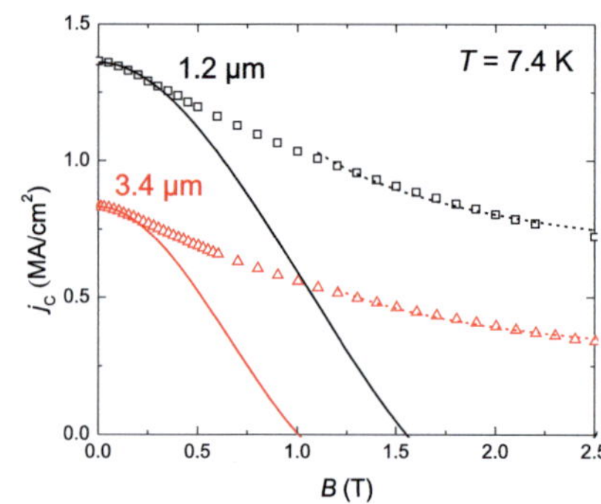

Figure 5.12: Dependence of j_C on external magnetic field of bridges structured from a Ba-122 thin film. Left: $j_C(B)$ of a narrow bridge at 12.7 K. The initial decrease is linear (solid line) and deviates from this behavior at higher B (dashed line is to guide the eye). Right: $j_C(B)$ for two larger bridges at 7.4 K. The solid lines are fits of eq. 5.8 to the low-B data, whereas the dashed lines are fits of eq. 5.9 to the high-B data.

that is too low to allow vortex entry across the penetration barrier. Consequently, the penetration current is reached only at much lower relative temperatures $T < 0.4 T_C$. A superconducting wire made from Ba-122 can thus reach the de-pairing critical current in a wide temperature range even in µm sized bridges.

Magnetic field dependence of the critical current density

Similar to the measurements on Nb, the reduction in $j_C(T)$ due to the width can be achieved by the application of an external magnetic field. The magnitude of the reduction however differs, depending on w and the temperature range. The dependence of j_C on the external magnetic field perpendicular to the sample surface is measured at a fixed temperature in a cryocooler. For all measurements, the dependence was non-hysteretic and symmetric to the polarity of the magnetic field. Figure 5.12 (left) shows the measured $j_C(B)$ dependence for a narrow bridge $w = 0.9$ µm for a relatively high temperature of $T = 12.7$ K. This is the regime where the de-pairing critical current is reached. The behavior of $j_C(B)$ fits well to the expected dependence described in section 5.1.1: For small fields the reduction of the critical current is linear (solid line). Here, the film is in the Meißner state and the critical state is reached once the first vortex enters into the film, since it is immediately swept to the other side of the bridge. For higher fields the dependence changes and follows the $j_C(B) \propto 1/B$ relation. However, the situation changes for wider bridges and for lower temperatures. Figure 5.12 (right) shows two measurements on bridges with $w = 1.2$ and 3.4 µm at $T = 7.4$ K. Instead of the relatively strong linear decrease of j_C for low fields, the reduction is now less pronounced and follows an approximately parabolic dependence.

The reason for the non-linear dependence is that the description in section 5.1.1 is only valid for the case of low bulk pinning materials. This assumption does not seem to hold for the Ba-

122 films. A theoretical description including non-neglectable pinning was given for the case $d \ll w \leq \lambda_{\text{eff}}$ [141], which is applicable to the Ba-122 bridges in the low-temperature range. For low fields, the critical current density is predicted as

$$j_C(B) = j_C(0)(1 - 6h^2 + 3\sqrt{3}h^3), \tag{5.8}$$

where $h = \pi \xi w B / \Phi_0 \mu_0$ is the magnetic field scale. Indeed the model fits well to the experimental data as shown by the solid lines in Fig. 5.12 (right). For large fields the dependence should change into

$$j_C(B) = j_C^p + \frac{2\pi \lambda_{\text{eff}} \mu_0}{c} \frac{(j_C^d - j_C^p)^2}{2wB}, \tag{5.9}$$

where j_C^p is the de-pinning current density. In the limit of no pinning or very large magnetic field, this dependence becomes the general $j_C(B) \propto 1/B$ form. Respective fits to the high field part of the measurements are shown as dashed lines in Fig. 5.12 (right). In order to quantitatively analyze the $j_C(B)$ behavior and its implications for pinning forces of the Ba-122 film, more systematic studies for different widths and temperatures have to be performed. However, from the observations described here we can conclude, that significant pinning forces are present in the Ba-122 films. This changes the current transport behavior of bridges structured from Ba-122 and allows relatively robust $j_C(B)$ dependences: For example, in the 3.4 μm wide bridge a magnetic field of 0.5 T only reduces the critical current to 83% of $j_C(0)$. The narrower the bridges and the higher the temperature, the stronger the reduction becomes.

To summarize, materials with low intrinsic critical current carrying ability suffer less from the self-field induced reduction of j_C below the de-pairing critical current. Although the absolute value of j_C is then lower, for applications that benefit from a high j_C/j_C^d ratio this can be a considerable advantage. It increases the temperature range of operation and relaxes the requirements of structuring in nano-scaled dimensions. Also, the magnetic field dependence reveal relatively strong pinning forces which make the critical current robust to reduction by external magnetic fields. This allows the consideration of Ba-122 as a material even for cases with challenging technology for low-dimensional structuring or for cases where a larger wire cross-section would be desirable.

5.1.4 NbN and TaN bridges

The previous sections, two superconducting materials have been studied that show quite different properties of the current transport in a superconducting bridge. To study the relevancy of the discussed effects on the superconducting nanowire detectors, we come back to common SNSPD materials. We first compare the observations with previous measurements on NbN, a material that was already introduced in detail in chapter 3. After that, the study is extended to TaN, a material very similar to NbN, which is also a nitride of a transition metal that becomes superconducting at low temperatures. Due to its lower critical temperature it has recently been considered for application in SNSPD and demonstrated excellent performance [36].

The temperature dependence of j_C for μm and sub-μm sized bridges made from NbN on sapphire was found to be rather similar to the Nb case [142]. All samples coincide close to T_C and follow eq. 5.3 well. At lower temperatures, j_C of the μm wide bridges deviates from this dependence and is reduced below the de-pairing values. The smaller the bridge width and the thinner the film thickness, the wider the temperature interval of de-pairing-limited current transport extends into the low temperature range. For very narrow bridges of 300 nm width, j_C^d could be retained over the full temperature range of measurement. A second series of bridges structured from a NbN film deposited on a Si substrate showed only a weak dependence on the bridge width. This indicates that the choice of substrate, and/ or the material properties of the NbN film grown on top of this substrate contribute to the current transport properties of the bridge.

Measurements of NbN in magnetic field [143] show that an external magnetic field has principally the same effect as increasing the bridge width, similar to what has been shown for Nb (compare e.g. Fig. 5.5). Fields in the range of several mT already lead to a significant reduction of j_C. Remarkably, the $j_C(T)$ dependence of some NbN samples was non-monotonous in the temperature range where the vortex penetration limits the critical current. To correctly model those dependences, a full description of vortex penetration, pinning and movement inside the bridge in the presence of applied current and external magnetic field would be required.

The critical current of bridges structured from a thin TaN film with widths between 60 nm and 8 μm was studied in [144]. Once again, essentially the same dependence as described for the Nb bridges can be observed. The smaller bridges follow the temperature dependence of j_C^d over almost the full temperature range. However, when w approaches the sub-μm range, the de-pairing critical current density also seems to be reduced. The $j_C(T)$ data then do no longer run together at $T \to T_C$, but follow the temperature-dependence $(1 - T/T_C)^{1.5}$ with a different pre-factor. This pre-factor is lower than the prediction by eq. 5.3, calculated with the respective material parameters, even if the reduction of T_C due to the proximity effect is accounted for. However, it is still assumed that it describes the de-pairing limit in the nanowire.

As a final argument for the described models of current transport in a nanowire, we conclude this section by looking at the dependence of the critical current on an external magnetic field at a fixed temperature of 4.2 K. The measurements are conducted on several bridges with width $w = 190$ nm to 5 μm, structured from a TaN thin film. Since the $j_C(B)$-dependences showed to be very sensitive even for very small magnetic fields, the measurements were conducted in a different setup than for the Ba-122 samples described above. The results are presented in Fig. 5.13 (left) for four representative bridges. Since the data are symmetric for both the polarity of B and I, only one quadrant of the measurement is shown. For direct comparison the data are normalized to the respective zero-field values. The initial linear decrease of $I_C(B)$ is observed as expected from the descriptions presented in section 5.1.1. The solid lines show linear fits to the $B \to 0$ data from which the slope dI_C/dB is extracted. The corresponding slopes are

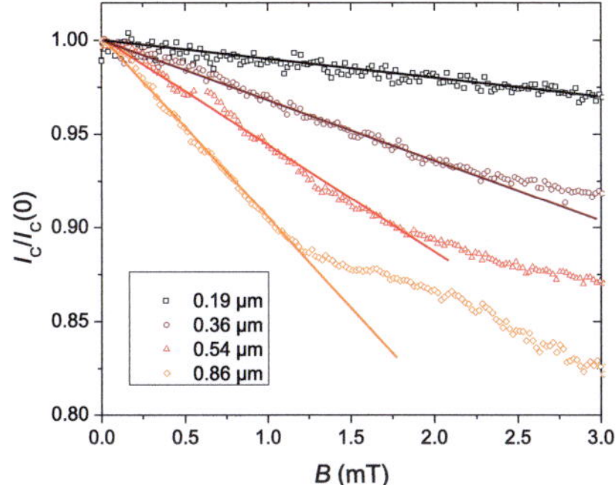
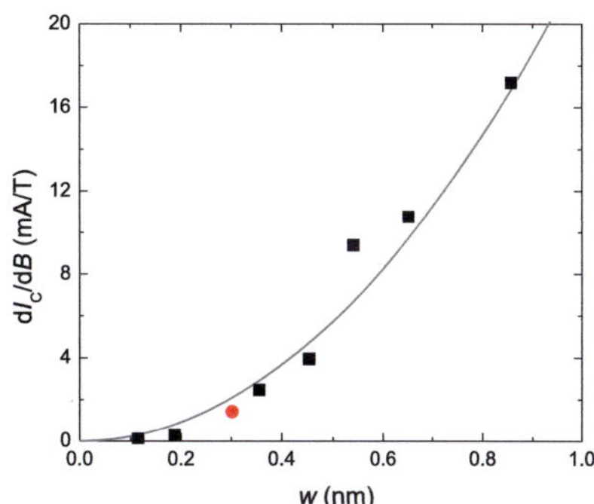

Figure 5.13: Left: Critical current in magnetic field B normalized to zero-field critical current for TaN bridges of different width. The solid lines are linear fits to the low-field data from which the slope dI_C/dB is extracted. Right: dI_C/dB values in dependence of the bridge width, where the red point marks a sample that was structured from NbN. The solid line is a fit of eq. 5.6 to the data.

shown in Fig. 5.13 (right) over the width of the bridge for samples up to $w \approx 1$ µm. Indeed the parabolic dependence on the sample width predicted for very narrow bridges by eq. 5.6 can be confirmed. For wider samples, the magnetic field measurements started to become hysteretic and non-monotonous as additional unsystematic features start to emerge, making an extraction of dI_C/dB impossible. The same measurement was conducted on a 300 nm wide NbN bridge, which showed almost the same dependence and is marked by the red data point. We can thus safely assume that these results are also directly applicable to the NbN case.

In conclusion, we found that the models presented in section 5.1.1 describe the situation in NbN and TaN nanowires fairly well. Thus, the mechanisms of vortex penetration over the edge barrier is the defining property of the experimental critical current in the samples. For the narrowest bridges, imperfections of the edge structure caused by the structuring process or oxidation of the surface layers become significant and may reduce the critical current of the bridges additionally. The patterning of nanowires without such contributions can be technologically challenging. For ideal film and edge quality, however, it should be possible to reach the de-pairing critical current at the operation temperature 4.2 K for nanowires of typical dimensions ($d < 5$ nm, $w \approx 100$ nm) used for SNSPD devices.

5.2 Geometry-dependent critical current in nanowires

In SNSPD detector structures, even for samples of high quality and very narrow width, the critical current density may still be significantly below the de-pairing limit. This is caused by the geometrical layout of the detector. In the most commonly used structure, the meander, there are sharp 180° turns at the end of each line as a connection to the next line. At the inner edge of these turns, the current density is increased and exceeds the current density of the lines. Therefore the critical current density is reached in the turns at a lower bias current than in the lines. This leads to a smaller measured critical current of the whole device with respect to a straight wire. Indeed a dependence of the critical current of SNSPD samples on the fill factor has been observed [13]. In this work, the fill factor of NbN detectors was increased by a continuous reduction of the gap between two nanowire lines, down to a minimum of 12 nm. The smaller gap leads directly to a sharper 180° turn at the end of the lines which suppresses the critical current of the device.

5.2.1 The current crowding effect

In normal metals, this effect is well known and called current crowding. It describes the concentration of the current density at the inner corner of a curve for a thin film structure with a width much below the skin depth $w \ll \delta$. In this case the effect adds a resistance on top of the geometrically expected resistance and attracted a lot of attention as the technology of the semiconductor industry advanced into the nanometer regime.

In the case of a superconducting nanowire, the effect reduces the effective critical current of the structure and the relevant length scales are the penetration depth λ and the coherence length ξ. The qualitative effect was already suggested in 1963 [145], but was only recently extended [43] to allow quantitative predictions. The authors consider several simple geometrical shapes and present the results on numerical calculations based on the London model. They also introduce the new aspect of a local modification of the edge barrier for vortex penetration by a corner, and define the critical current as the current for which first nucleation of vortices is possible.

In typical detector structures, $\xi \ll w \ll \lambda_{\mathrm{eff}}$ can be assumed and the current density in the straight lines is homogeneous. If a nanowire follows a curve with constant width, the current density is condensed at the inner edge of the curve and the penetration barrier is minimal at the minimum radius of curvature (see Fig. 5.14). For the case of a 90° turn with sharp corners and width w, the calculations yield a reduction factor R of the critical current with respect to I_C of a straight wire of the same dimensions of

$$R = \frac{I_C(90°)}{I_C(0°)} = \frac{3}{2}\left(\frac{\pi\xi}{4w}\right)^{\frac{1}{3}}. \tag{5.10}$$

For the case relevant for a SNSPD meander design, a 180° turn with sharp corners, the exact reduction factor depends on the width of the gap between the lines but is in all cases $R \propto (\xi/w)^{1/2}$ [43]. This reduction describes the maximum achievable critical current at $T = 0\,\mathrm{K}$. In an experimental situation, thermal excitation of vortices across the barrier can cause the transition to the normal conducting state already at lower currents and leads to a statistical distribution of measured critical current values.

The effect of current crowding can lead to the peculiar situation, that the application of an external magnetic field effectively increases the critical current of a nanowire. This can occur in

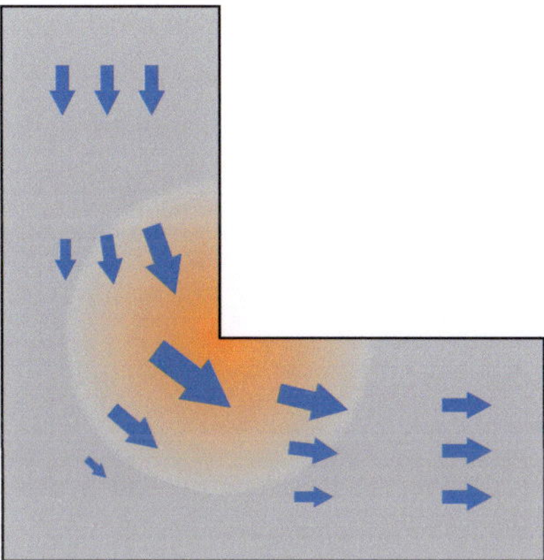

Figure 5.14: Schematic illustration of current accumulation in a nanowire with a rectangular bend. The arrows symbolize the direction of current flow and their size the local current density. At the inner edge of the bend (orange area), the current density is increased above the straight line value.

structures with a single bend, when current crowding at the inner edge reduces the critical current below the value of a straight wire. The Meissner currents induced by an applied magnetic field are at one edge of the wire parallel to the bias current and at the other edge opposite. If the induced current is on the inner edge of the bend opposite to the current, the current density is reduced and the current crowding effect is reduced. In the other case the induced current leads to a further increase of the current density and the current crowding effect is amplified. Thus, the critical current of the structure is either increased or decreased, depending on the direction of the applied field. Consequently, the dependence $I_C(B)$ is expected to become asymmetric with the polarity of B. The maximum critical current is reached at a field for which the current crowding effect is negated. This value, however, is still below the critical current of a straight nanowire at $B = 0$.

Numerical calculations on the basis of the London equations for this situation were recently presented [146]. The predicted $I_C(B)$ dependence is similar to the straight wire case (Fig. 5.1), but with a reduced maximum that is shifted on the B-axis. Simulations of the time-dependent Ginzburg-Landau equations confirm the assumption, that the current crowding leads to a local reduction of the barrier for vortex penetration and that the vortices (or antivortices) enter the wire at the point of minimum curvature once the critical current is reached and are then swept by the Lorentz force to the opposite edge. For a 90° turn with $w = 15.5\xi$ and typical Nb thin film parameters, an increase of I_C with magnetic field by 34% is predicted.

In summary, current crowding is an effect that has been proposed to have significant influence on the critical current of structures with dimensions and shapes typical for nanowire detectors.

Not only this, also the distribution of the current is affected, which can lead to strongly locally different I_B/I_C^d ratios. Theoretical descriptions show that the expected effect should scale with the ratio ξ/w and thus becomes less influential as the width of the nanowire is reduced. To what extend the predictions can be applied to real detector structures is the topic of the experimental investigations described in the next sections.

5.2.2 NbN nanowires with bends

To systematically study the reduction of the critical current density by geometry, instead of measuring on detector samples which can have rather complicated layouts, we look at a simple bridge structure. This case was discussed in the last section, and it was found that for samples with $w \ll \lambda_{\text{eff}}$, the current distribution is homogeneous across the wire cross-section and that the depairing critical current density, at least in a sample with ideal edges, can be achieved. In this nanowire layout a bend is introduced which has a certain angle α (defined as the angle between the straight continuation of the wire and the new wire path) with a certain bending radius r (defined by a circle segment along the inner edge of the bend). We first evaluate which sample types and parameter ranges are of interest by a simple simulation. Based on this, samples were designed, structured and the critical currents measured for the experimental verification of this effect. The results of these measurements and their discussion were published in [45].

Maxwell simulations

A simple simulation serves as a rough visualization of what can be expected from the current distribution in such a bend. Here, only the current distribution due to the pure geometry is considered, disregarding any superconducting properties, in analogy to a normal conductor with a width smaller than the skin depth $w < \delta$. The problem then reduces to the solution of the Maxwell's equations by a FDTD-solver. As a model we assume a two-dimensional structure, which is well justified in the case of a thin film, with a homogeneous constant resistivity. The width of the structure is always 100 nm, with a current flowing in and out at the boundaries. The simulations were conducted with COMSOL multiphysics v4.1. More details are given in [147].

Figure 5.15 shows the current distribution in bends with angles $15°$, $45°$ and $90°$, respectively, with a bend radius of 40 nm. The color refers to the current density with respect to the density at the straight parts (green): blue colors symbolize areas of reduced current density, while red colors are areas with increased current density. Even for this simple model, the current 'crowding' at the inner edge of a bend is apparent. The maximum value of the current density is in all cases located at the middle of the inner edge of the bend and increases with the angle. Also, the inhomogeneous part of the current distribution extends farther into the center of the wire as the angle increases. On the other hand, this effect seems to be reduced as the radius of the bend is increased. Figure 5.16 (left) shows the same $90°$ bend with $r = 40$ nm. When the radius is increased (middle: 100 nm, right: 500 nm) while the width of the wire being kept constant, the

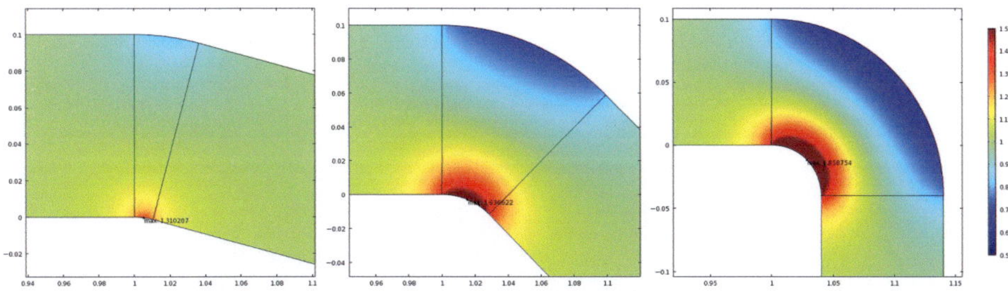

Figure 5.15: Simulated current distribution of 100 nm wide nanowires with bend angle α 15° (left), 45° (center) and 90° (right). The color represents the current density with respect to the straight line from lower (blue) to higher (red). A minimum bend radius of 40 nm is assumed to reflect the resolution of a lithography process.

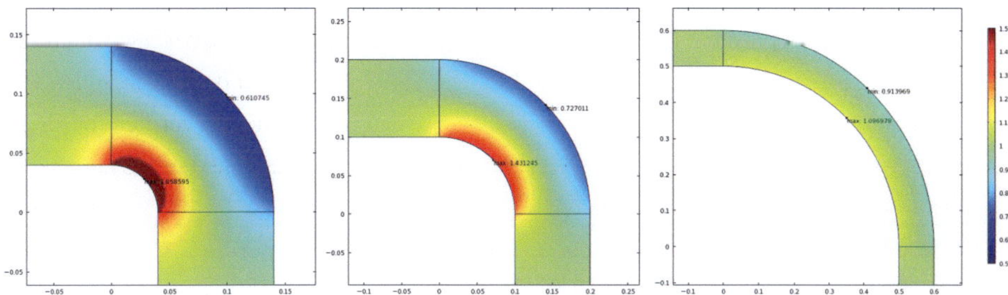

Figure 5.16: Simulated current distribution of 100 nm wide nanowires with 90° bend angle for inner bend radii r of 40 nm (left), 100 nm (center) and 500 nm (right). The color represents the current density with respect to the straight line from lower (blue) to higher (red).

current crowding effect is reduced. Instead of being a single spike, the current density is now distributed smoothly along the whole length of the bend and is only slightly increased above the current density of the straight part. Also, the distribution across the nanowire becomes more homogeneous.

Finally, we look at a structure with dimensions of a meander turn in a typical SNSPD device: The nanowire has a width of 100 nm with a 180° turn and a 100 nm gap between the wires. The devices often have an elongated turn of about twice the nanowire width, which is supposed to reduce susceptibility of the turns to fabrication errors. Since in real devices, corners are always rounded due to resolution limitations of the employed lithographic process, a rounding radius of 40 nm is assumed. Figure 5.17 shows the current distribution simulation a meander edge. Again a strong current crowding is apparent at the inner edge of the turn. The distribution varies only slightly from two 90° turns in succession, which is due to the short straight part in the turn. However, for samples where the gap between the nanowires is smaller (e.g. to increase the fill

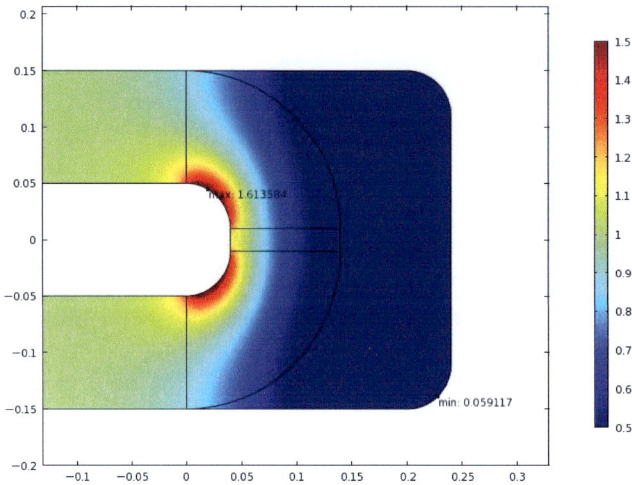

Figure 5.17: Simulated current distribution of a typical SNSPD 180° meander turn. The wire as well as the gap between wires are 100 nm wide. A minimum bend radius of 40 nm is assumed to reflect the resolution of a lithography process. The color represents the current density with respect to the straight line from lower (blue) to higher (red).

factor), the effect will be even stronger. We also note that the elongated part at the turn does not seem to reduce the problem, since this area carries almost no current.

The simulations suggest that the current crowding effect is indeed present in SNSPD meander devices and causes a strong increase of the local current density at the inner edge of a turn. This limits the overall device critical current, since the critical state is reached already when the current density in this region surpasses the critical current. Also, the non-homogeneity of the current distribution and consequently of the ratio I_B/I_C leads to a strongly varying local detection efficiency. According to the simulations, the effect should be reduced for bends with smaller angles and larger radii. However, it has to be emphasized that for direct comparison with experimental values, simulations have to be made on the basis of the London or Ginzburg-Landau equations to include properties unique to superconductors (for example the penetration of magnetic vortices).

Experimental procedure

With this preconsiderations in mind, two series of nanowires were structured from one NbN thin film deposited on a sapphire substrate. In each case the wire had two identical bends (see Fig. 5.18 for the principle layout). In the first series the bends were abrupt with an angle α varying between $0°$ and $90°$. In the second series, the angle $\alpha = 90°$ was fixed but the bend was structured with an arc segment of a circle, with the inner radius r and the outer radius $r + w$. r was

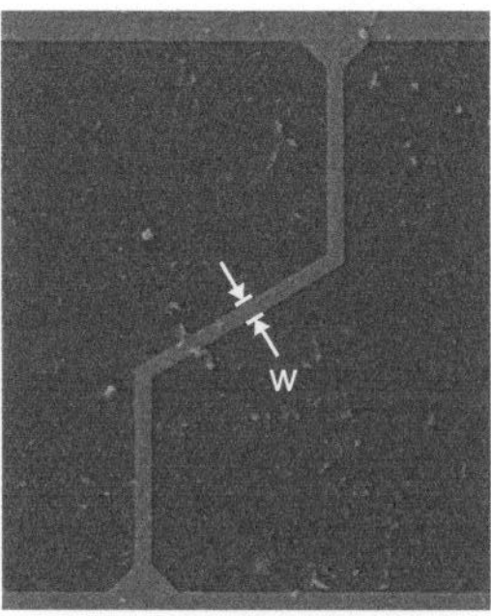

Figure 5.18: SEM image of a nanowire with two sharp bends. The light part is the NbN film and the dark part the bare substrate surface. The horizontal parts at the top and the bottom are the edges of large contact pads for wire bonding to which the nanowire is contacted by tapered connections. The width of the displayed wire is 300 nm.

varying up to 500 nm. For the thickness of the film $d = 10$ nm and the width $w = 300$ nm were selected. For those dimensions, the bridge is well in the limit $w \ll \lambda_{\text{off}}$ as shown in section 5.1.4. On the other hand the ratio w/ξ is relatively large, so that a strong current reduction effect can be expected. To avoid any additional current crowding, the connections of the wire to the contact pads are tapered. The nanowire was structured by electron-beam lithography with a resolution <5 nm and a consecutive reactive ion-etching process. The contact pads were defined by photolithography and etched by the same RIE process. The exact sample thickness was determined after the structuring process with a profilometer to be 10.4 nm. The actual nanowire widths were evaluated by scanning electron microscopy with an accuracy <10 nm. Those values were then used for the calculation of the critical current density $j_C = I_C/wd$, where the inaccuracy of the width determination introduces the largest part of the error.

To verify that the structuring of the bend did not influence on the overall superconducting properties but only on the critical current density, the critical temperature was measured on all samples after the structuring process. The results (Fig. 5.19) show a spread of T_C values between 13.1 and 13.6 K, which is typical for 300 nm wide nanowires. No systematic dependence of T_C on either α nor r is observed.

The critical current densities measured at 4.2 K are shown in Fig. 5.20 for the sample series with varying α. For the straight reference sample ($\alpha = 0°$), the maximum value $j_C(4.2\,\text{K}) = 13.3\,\text{MA/cm}^2$ is achieved. With increasing angle, j_C is gradually reduced in accordance with expectations. For the 90° sample $j_C(4.2\,\text{K}) = 8.1\,\text{MA/cm}^2$, which is only $\approx 60\%$ of the straight wire. The corresponding values for the series with varying radius are shown in Fig. 5.21. The samples with $r = 50$ nm are the same as the $\alpha = 0°$ samples from the previous series. With increasing r, the $j_C(4.2\,\text{K})$-values once again increase, up to $10.7\,\text{MA/cm}^2$ for $r = 500$ nm. This is more than 80% of the straight reference sample's value.

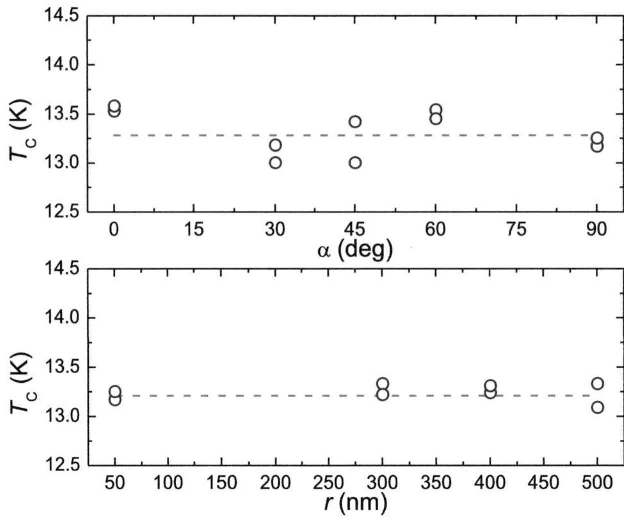

Figure 5.19: Critical temperature T_C of nanowires with bends of different angle α (top) and $90°$ bends with different radii r (bottom). The $0°$ samples correspond to straight lines. The $\alpha = 90°$ samples from the top are identical to the $r = 50$ nm samples from the bottom, which is the minimum achievable sample radius due to the limitations of the structuring process. The dashed lines are a guide for the eyes.

To investigate the mechanism for this reduction more closely, we look at the temperature-dependence of j_C between 4.2 K and T_C. Figure 5.22 shows the $j_C(T)$-dependences for three selected samples with angles $\alpha = 0°$, $45°$ and $90°$. These samples were selected for their smallest difference in the measured width to avoid additional impact of the sample width on $j_C(T)$ [77, 144]. The principle behavior is, however, similar for all samples: Close to T_C, the j_C values of all three samples practically coincide. At temperatures below $\approx 0.7\,T_C$, the j_C values deviate from each other. For the sample with the larger bend angle the j_C values stay lowest, whereas in the straight line they continue to rise with decreasing temperature almost undisturbed. The difference between the samples grows as the temperature decreases.

This is the same general dependence as was observed in section 5.1 for straight nanowires when the wire width was reduced. We can thus assume that for temperatures $T > 0.7T_C$ the de-pairing critical current is reached in all three nanowires and that the reduction of j_C at low temperatures is also in this the case caused by the penetration of magnetic vortices into the wire.

Comparison with theoretical predictions

To compare the results with theoretical predictions, we choose the sample with the sharp $90°$ bends, where the strongest reduction was observed. The critical current densities are normalized

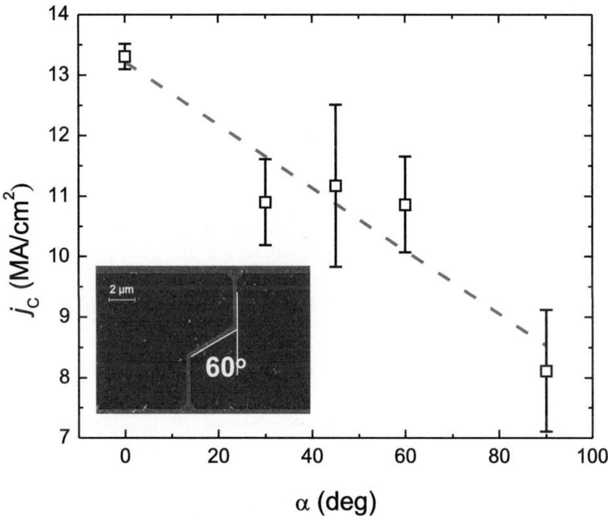

Figure 5.20: Average critical current density j_C of NbN nanowires with two bends of angle α. The sample layout and the definition of α are shown in the inset on an SEM image of one sample. The general trend of j_C reduction with increasing angle is symbolized by the dashed line.

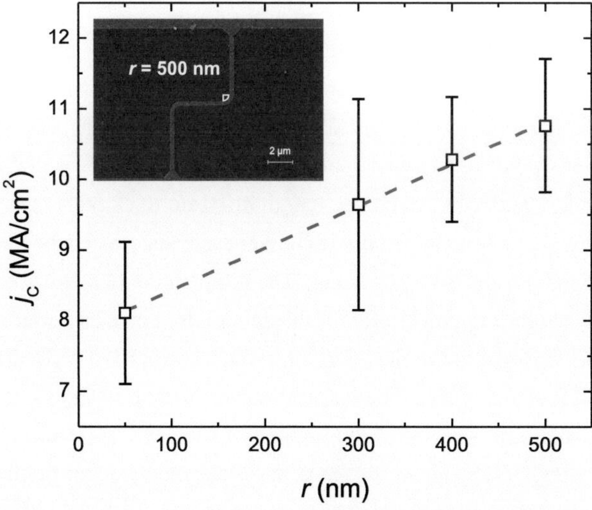

Figure 5.21: Average critical current density j_C of NbN nanowires with two 90° bends with radius r. The sample layout and the definition of r are shown in the inset on an SEM image of one sample. The general trend of larger j_C with increasing angle is symbolized by the dashed line.

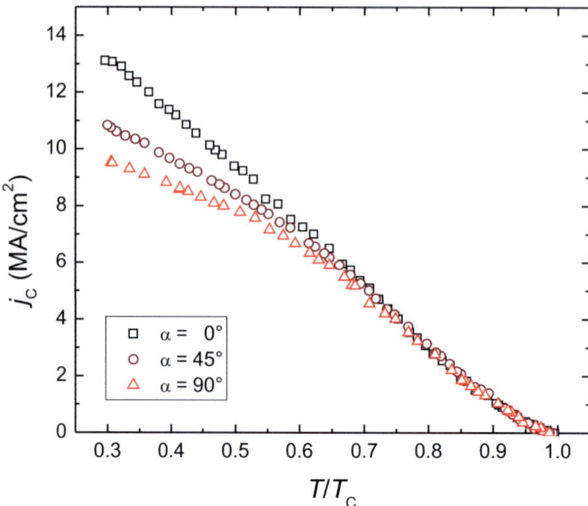

Figure 5.22: Current density j_C vs. reduced temperature T/T_C for three selected samples with $\alpha = 0°$, 45° and 90°. The samples coincide for temperatures close to T_C and diverge in the low temperature region.

to the respective values of the straight reference sample $j_C(90°)/j_C(0°)$, retaining only the contribution due to the geometrical influence. The respective temperature-dependence is shown in Fig. 5.23

Numerical calculations on the basis of the time-dependent Ginzburg-Landau equations [108] were supplied by Dr. D. Vodolazov of the Institute for the Physics of Microstructures from the Russian Academy of Sciences. The critical current of a superconducting line of rectangular cross-section and one sharp 90° bend was calculated for a linewidth of $w = 60\xi(0)$. This reflects the experimental situation, since the coherence length in 10 nm thick NbN is roughly 5 nm [19]. The model assumes that the sample is initially in the Meissner state which becomes unstable at a certain current, which defines the critical current. The temperature dependence of the reduction of the critical current, calculated numerically for this model, is shown as a solid line in Fig. 5.23.

Additionally, the reduction of the critical current in the framework the London equations was computed with Eq. 108 of ref. [43]. The respective dependence is shown as the dashed line in Fig. 5.23. Both calculations are limited to temperatures below $0.9T_C$, because the coherence length diverges for $T \rightarrow T_C$. They show the same basic dependence that predicts a decrease of the critical current for low temperatures. However, the values calculated by the GL approach are by a factor ≈ 1.3 larger than those of the London approach. The difference can be explained by the definition of the energy barrier for a single vortex entry. There is a free parameter, which formally describes the minimal distance at which vortex may approach the edge of a straight

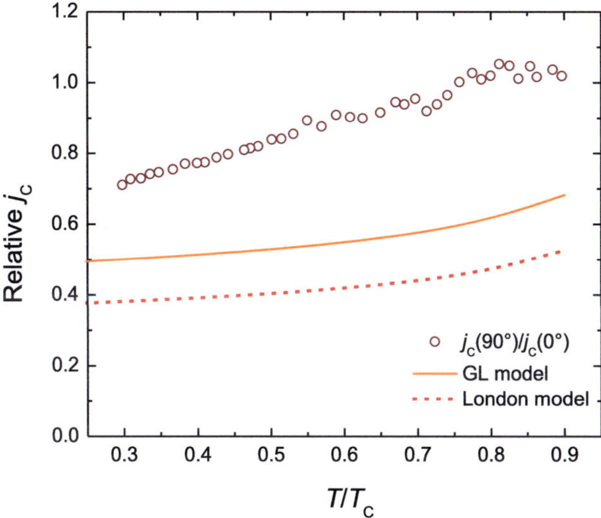

Figure 5.23: Experimental critical current density of a sample with a 90° bend normalized to the critical current density of the straight reference sample plotted over reduced temperature T/T_C. The lines shows the calculated current reduction on the basis of the TDGL equations (solid line) and the London model (dashed line).

line. For consistency the minimal distance is defined by setting the barrier to zero for the current equal to the GL de-pairing current. Although slightly different in different publications, this distance appears comparable with the size of the vortex core. On the other hand, the barrier in the London model can be defined only assuming that this distance is larger than the radius of the core. Calculations for a narrower line ($w = 20\xi(0)$) show that the GL and London critical currents differ by the same factor ≈ 1.3 only at $T \leq 0.6T_C$; at larger temperatures this factor decreases and depends on the ratio $w/\xi(T)$.

Even so, the critical current reduction found by the experiment is not as strong as any of the two models predicts. At the lowest temperature $0.3T_C$, the experimental ratio $j_C(90°)/j_C(0°)$ reaches 0.65, whereas the GL model predicts 0.5 and the London model only 0.38. Assuming that finite rounding is present even in nominally sharp 90° bends and using the expressions for the reduction factor from Ref. [43], we have found a negligibly small correction to the critical current.

The origin of the discrepancy between the theoretical calculations and the experimental data remain so far unclear. Other recent work [44] reports similar measurements on NbTiN structures. Here, the observed discrepancy between theory and experiments was similar. In a 90° bend of 60 nm inner radius, $R = 0.51$ was measured where $R \approx 0.33$ was expected from [43]. One possible explanation for the discrepancy that was put forward in [45] could be that both NbN

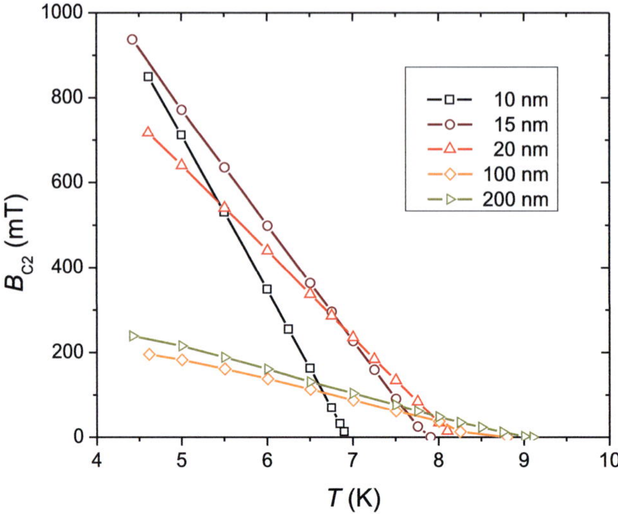

Figure 5.24: Dependence of upper critical magnetic field B_{C2} on temperature for unstructured hot-deposited Nb thin films of various thickness.

and NbTiN are granular materials. To verify this hypothesis, the experiment was repeated on hot-deposited Nb films. The results are presented in the next section.

5.2.3 Nb nanowires with bends

In contrast to NbN, Nb can be deposited into thin films with high uniformity. Like in NbN, the uniformity can be further improved when the substrate is heated during deposition. The parameters are strongly dependent on the thickness of the Nb films. To come to comparable measurements of the critical current reduction as for NbN, the structures should have similar ratios of ξ/w. Also, the thickness of the film should not be much larger than ξ to remain in the two-dimensional regime.

The Nb thin films were deposited by dc magnetron sputtering of a pure Nb target in an Ar atmosphere at a pressure of $p_{Ar} = 1.39 \cdot 10^{-3}$ mbar. The sapphire substrates were heated to 800°C and kept at this temperature during deposition. Details about the optimization of the deposition parameters and the characterization of the Nb thin films can be found in [148]. The key parameters of a series of films with thickness from 10 to 200 nm are compiled in table 5.1. The critical temperature T_C, the residual resistivity ratio RRR and the normal state resistivity ρ were determined by $R(T)$ measurements of the samples. The electron mean free path can be estimated from the resistivity with the help of the relation $l\rho = 3.72 \cdot 10^{-6}\,\mu\Omega\text{cm}^2$ [149], which is to good approximation constant in Nb. In the case of the 15 nm thick film, $l = 6.8$ nm.

Table 5.1: Parameters of unstructured thin Nb films extracted from resistive and magnetic measurements.

| d | T_C | RRR | ρ | $B_{C2}(0)$ | D | $\xi(0)$ |
nm	K		$\mu\Omega$cm	mT	cm^2/s	nm
10	7.06	2.8	11.2	1.81	2.95	13.5
15	8.09	4.1	5.5	1.52	4.04	14.7
20	8.50	5.1	3.9	1.19	5.39	16.6
100	9.42	26.1	0.5	0.32	22.29	32.3
200	9.67	58.2	0.2	0.37	19.67	29.7

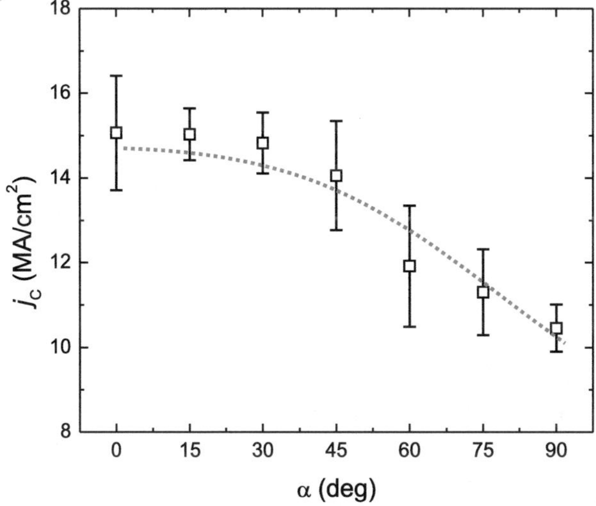

Figure 5.25: Dependence of the average critical current density of nanowires on the bend angle α. The samples were structured from the same 15 nm thick Nb films deposited on sapphire. The dashed line illustrates the general trend.

The other parameters are determined from the measurements of the upper critical magnetic field in dependence of the temperature $B_{C2}(T)$ analogous to the description in section 3.3. The measurements are shown in Fig. 5.24. In contrast to NbN, the slope $\mathrm{d}B_{C2}/\mathrm{d}T$ depends strongly on the thickness of the sample. Consequently the coherence length is changed significantly with the thickness. Also, the linearity of $B_{C2}(T)$ continues for the Nb samples all the way up to T_C, which is an indication that the material is less granular.

From these measurements, we find that the condition of $\xi \approx d$ is given for all films between 10 and 20 nm. The film with $d = 15$ nm is used to pattern sample structures. The width of the nanowires is set to 600 nm, giving a ratio $\xi/w \approx 0.025$, comparable to the NbN structures. Besides the change in the width, the design of the structures remain unchanged (compare Fig. 5.18) and the identical structuring process was employed. For each specific angle, 3 to 4 structures were made. The final samples had a T_C that was reduced by approximately 1 K, but with no dependence on the angle. The critical current measurement, however, revealed the reduction with increasing α. Figure 5.25 shows the average j_C values for the samples measured at 4.2 K. The dashed line illustrates the general trend of reduction of j_C with increasing angles. Since there are more α values than in the NbN experiment, the dependence is better resolved and can be recognized as approximately sinusoidal instead of linear.

The critical current density of the straight, $0°$ reference sample is 15.1 MA/cm^2. For the $90°$ sample, j_C is 10.5 MA/cm^2, which is 67% of the straight bridge. Remarkably, this result is almost the same as for NbN, despite the fact that the material parameters of the Nb films are significantly different. We thus conclude that the material properties do not influence strongly on the current crowding effect if the ratio ξ/w of the nanowires is similar.

5.2.4 Magnetic field enhancement of critical current

In this final section of the chapter, first experimental results regarding the enhancement of the critical current by external magnetic field, as predicted in [146], are presented. The effect should arise in nanowires with bends, if the screening currents introduced by the magnetic field oppose the current flow at the inner edge and thus reduce the current crowding effect (see section 5.2.1). The samples used in the previous measurements are not suitable for this investigation, because they have two turns with opposite bend direction (see e.g. Fig. 5.18). This means that a magnetic field normal to the sample surface will induce currents that have opposite direction at the inner side of each turn. An applied bias current will then be always parallel to the induced currents in one turn and anti-parallel in the other turn, independent of the polarity of bias current or magnetic field. Thus the current crowding is reduced in one turn but increased in the other turn and no net improvement is gained.

To observe the reduction of the current crowding by magnetic field, the sample must have only left-handed or only right-handed turns. In the simplest case, of course, the nanowire has only one turn. Since the contact pads of the photomask are positioned opposite of each other, this is not entirely possible. However, if the nanowire is first tapered to a much larger width and then connected to the pads with large radius segments, the contribution of these part should be neglectable. SEM images of bridges made with such a new design are shown in Fig. 5.26. The design width of the nanowires are in each case 600 nm wide and 2 μm long before it is widened by an approximately $45°$ stub. It is then connected to the contact pads seen in the top and bottom of the images, where any sharp turns are omitted. The bend angle α is defined as illustrated in the right image, so that the straight bridge corresponds to $\alpha = 0°$. The bridges are patterned from a 15 nm thick Nb film very similar to the one described in the last section. After structuring the actual width was controlled by SEM measurements and found to be only $\approx$500 nm, however almost equal for all bridges. Detailed parameter characterization of all samples can be found in [148].

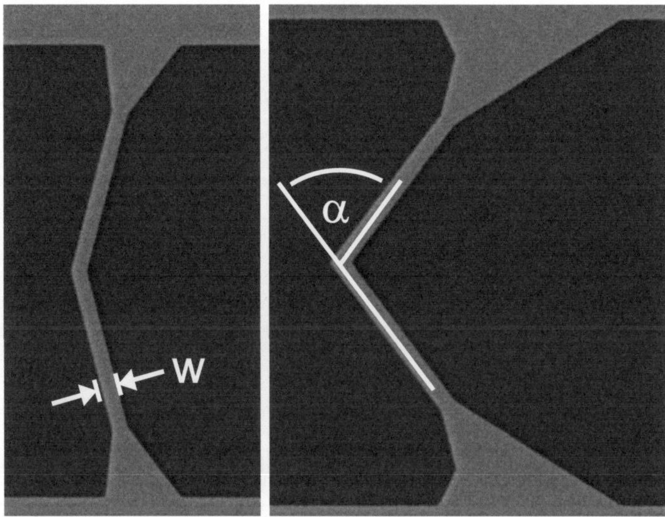

Figure 5.26: SEM image of $w = 600\,\mathrm{nm}$ wide nanowire samples made from 15 nm thick Nb films. Two samples with different bend angle α, defined as illustrated in the right image, are shown. The horizontal parts in the top and bottom are the contact pads.

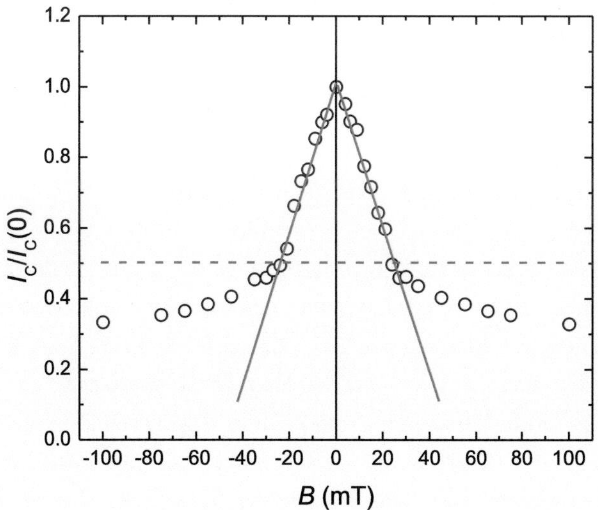

Figure 5.27: Critical current vs. external magnetic field applied perpendicular to the sample surface, normalized to the zero-field critical current $I_C(B)/I_C(0)$, for a 500 nm wide Nb bridge without bends. The dependence is symmetric to the magnetic field and applied current polarity. The reduction of $I_C(B)$ for $|B| \to 0$ is linear (solid lines) down to about $0.5 I_C(0)$. The measurements were performed at 4.2 K.

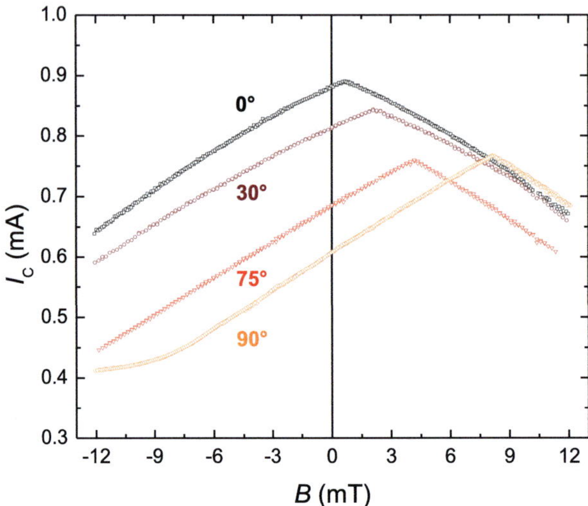

Figure 5.28: Critical current dependence on external magnetic field measured on four Nb bridges with different α. The same dependence is attained if both current and magnetic field polarity are reversed simultaneously. The measurements were taken at 4.2 K.

Measurements of the critical current in dependence on external magnetic field at 4.2 K are shown for a straight bridge in Fig. 5.27. The critical current is normalized to the maximum critical current at $B = 0$. Indeed, the dependence is symmetric for both magnetic field and bias current. The first decrease of I_C is approximately linear (solid lines). At about 0.5 I_C, marked by the dashed line, the dependence changes as the static vortex state is reached. The overall $I_C(B)$ dependence reflects very well the model expectations (compare Fig. 5.1). The measurements exceeded the shown range of ± 100 mT, but j_C continues to drop in this range with the same dependence. The general behavior for higher magnetic fields outside the linear regime was the same as for the 0° bridge for all samples. We thus compare in the following the measurement conducted in a second experimental setup, where a higher resolution in a smaller magnetic field range can be achieved. Additionally, we note that the $I_C(B)$ measurements are very sensitive to the individual samples. Although samples with obvious defects found in the SEM inspection or irregular values in the $R(T)$-characterization were already sorted out and not measured, there were still many samples where the $I_C(B)$ displayed hysteretic parts or discontinuous jumps. In the following only the samples with reproducible, non-hysteretic and continuous $I_C(B)$ dependences are considered.

Results for samples with differenet bend angle are shown in Fig. 5.28 for the positive critical current. The same dependence is measured for the negative critical current when the magnet field polarity is reversed. Due to the better resolution we can see now that the straight bridge (black) is not exactly symmetric, but has a slightly shifted maximum of I_C. This was observed for all $0°$ samples, where shifts both to positive and negative fields are found. This is probably due to the non-ideal edge fabrication that leads to some minor asymmetry between the two sides. Also, the reduction with increasing B is not perfectly linear, but slightly curved. For $B = 0$, the standard reduction of the critical current with increasing bend angle α is found. However, when considering the whole $I_C(B)$, the shift of the maximum along the B-axis with increasing α can be seen. Although the maximum value of I_C is in all cases below the one of the straight bridge, it is still considerably higher than the value at $B = 0$. For the $90°$ sample shown in the figure, an increase of the critical current by the applied magnetic field of 25.8% was achieved, which was the highest ratio of all measured samples.

The measurements agree qualitatively well with the theoretical predictions presented in [146] for a $90°$ turn: The maximum value of I_C should be reduced with respect to a line without turns and shifted to either side (depending on whether it is a left-handed or right-handed turn relative to B). Further simulations of the time-dependent Ginzburg-Landau equations for a $90°$ turn with film parameters typical for Nb are very similar to our measurements. The dome around the maximum is here also not linear but even stronger rounded. The explanation given is that the simulations were made for a width $w = 15.5\xi$, where the limit $\xi \ll w$ is not strictly valid anymore. In our case the wire is wider, approximately $w = 35\xi$, so that the rounding should be less pronounced, which is exactly what is observed. Due to the difference in width we can not directly compare the measurement values with the simulations, but the predicted shift in the maximum of I_C and the correspondent increase of I_C with the magnetic field can be clearly observed. It should also be noted that at the transition to the vortex state, we observe an increased statistical fluctuation of the I_C measurements, which could very well be understood by the vortex dynamics shown in the simulations, considering the non-ideal sample edges and non-zero temperature in the experiment.

The simulations were only presented for a $90°$ and a $180°$ turn in [146], but for the latter a stronger increase of I_C with magnetic field was predicted. To investigate the dependence on the bend angle α, for each sample the ratio of the maximum critical current $I_{C,max}$ to the critical current at zero magnetic field $I_C(0)$ is evaluated. The results are shown in Fig. 5.29. Up to $\alpha = 30°$, no significant improvement of $I_{C,max}/I_C(0)$ with angle can be observed. For larger angles, there is an upwards trend illustrated by the dashed line up to an average value of 1.15 for the $90°$ samples. Due to the high sensitivity of the experiment to slight irregularities in the edge shape of the bridges, a large spread in the data is observed.

Once again, the increase is not as strong as expected from theoretical models. It should be remembered, however, that already the initial reduction of I_C by the current crowding effect was not as strong as expected, and the application of the magnetic field is supposed to counter the effects of current crowding. In this way, the results are not surprising. In any case they show that the effect is measurable and that it could in principle be applied to an SNSPD to increase its critical current. For this, the detector layout should only have turns in one direction, which

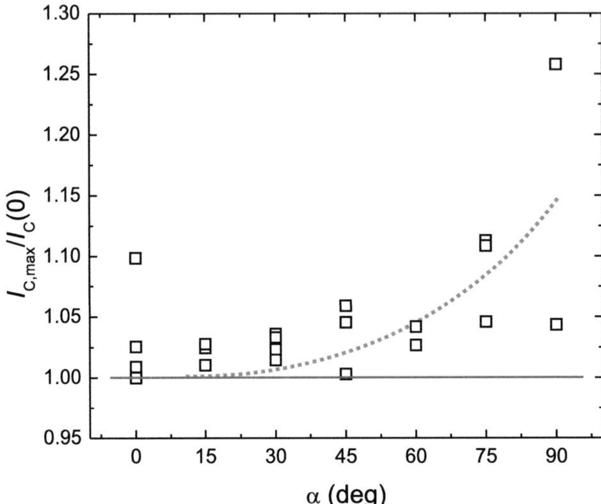

Figure 5.29: Ratio of maximum critical current achieved with applied magnetic field to the critical current without magnetic field in dependence on the angle α of the samples. The solid line shows a ratio of 1, meaning the maximum is at zero field. The dashed line illustrates the increase of the ratio for higher α.

means it will be some kind of spiral going only inwards. The connection for the bias current would then require an additional structure layer, making it technologically more challenging.

5.3 Summary

In this chapter, the critical current in a nanowire was investigated experimentally. For this special case of dimensions $d < \xi$ and $w \ll \lambda_{\text{eff}}$, the current density is distributed uniformly over the wire cross-section. The experimentally observed critical current is caused either by the de-pairing of Cooper pairs, which marks the ultimate limit of the nanowire critical current, j_C^d, or by a premature switching into the normalconducting state due to vortex penetration and movement. In the latter mechanism, the vortices have to overcome a penetration barrier at the nanowire edge. A quantitative description of the edge barrier height, taking into account the the vortex deformation energy at an edge of non-ideal shape and roughness, remains an open problem.

In a straight nanowire, j_C^d can be reached in a wide range of T_{dev} to T_C. The range is increased as the nanowire cross-section decreases, i.e. for small width or film thickness, giving an additional incentive to structure nanowire detectors as small as possible. This is due to the lower self-field that is created by the bias current through a narrower wire, which lowers the penetration barrier for the vortices in the same way as an external magnetic field. Another approach to improve the j_C^d range is the use of materials with lower intrinsic critical current density, so that de-pairing remains the limiting factor before the currents causing high self-fields are reached. Additionally, materials with higher vortex penetration barriers or with increased bulk pinning to impair vortex movement inside the film make the nanowire less susceptible to the self-field. With such material properties, it would be possible to operate a SNSPD close to j_C^d even for applications that require a relatively large cross-sections. For example SNSPD employed for the detection of X-rays require large film thicknesses to achieve feasible absorptance values [8].

For a NbN nanowire of commom SNSPD dimensions, however, ratios j_C^m/j_C^d at 4.2 K are already very close to one for a straight line with ideal edge quality. Nevertheless, in a standard meander-shaped detector layout, current crowding at the turns leads to an inhomogeneous distribution of the current density and reduces j_C^m further. The effect can be reduced by increasing the bend angle or radius and scales with ξ/w, but seems to be only weakly dependent on material. The experimental investigations verified the theoretical predictions qualitatively, albeit with lower reduction factors. The reason for this discrepancy remains so far unclear. However, it was also verified experimentally, that the current crowding effect can be countered by the application of an external field, effectively increasing the critical current of the nanowire. This can only be realized in a structure with turns exclusively in one direction, which for a larger area coverage would neccessitate an additional wiring layer to make contact. It might also remain an option for a waveguide-coupled SNSPD, where two long parallel wires with only one turn as connection are required.

To improve the performance of an SNSPD, the two most important aspects concerning the critical current are:

- To achieve a critical current that is as close to the de-pairing limit as possible ($I_C^m/I_C^d \approx 1$). This will lead to a large cut-off wavelength, because $\lambda_C \propto (1 - I_B/I_C^d)^{-1}$ and consequently to a high detection efficiency in the infrared range.

- To homogenize the local distribution of the I_C^m/I_C^d ratio, so that a high detection efficiency can be reached in all parts of the structure instead of being limited to small areas.

To this end, film thickness and wire width should be as small as possible while still allowing for the patterning with high uniformity and edge quality. Turns, which are neccessary to fill a certain detector area with densely spaced nanowires, should have low angles and/ or large radii. Connections to the external bias lines should be not aprupt, but tapered to larger widths. A detector layout based on a spiral shape is proposed as an alternative to the standard meander layout and will be presented and discussed further in chapter 6.

6 Spectral detection efficiency of SNSPDs

In the previous chapters, the theoretical description for the performance of SNSPDs in the framework of the hot-spot model was discussed. The NbN thin film material properties, the energy relaxation processes in the films and the critical current density of nanowires were investigated in detail. On the basis of the obtained results, approaches to improve SNSPD devices were formulated. The performance of SNSPD is defined by four main properties:

- The first and most important parameter is the detection efficiency. It defines the probability that an incoming photon will cause a count event registered by the detector and reflects the sensitivity of the device.

- The second parameter is the rate at which count events are recorded even though no photon is incident on the detector, the so-called dark count rate (*DCR*). This parameter is equivalent to the noise of the device.

- Third, the so-called cut-off wavelength describes the extend of the detection efficiency's spectral dependence into the infrared range.

- Finally, the maximum count rate is limited by the relaxation time constant of a response pulse.

For the direct experimental evaluation of those detector properties, SNSPD devices are patterned from the NbN thin films with a combination of electron-beam and photolithography. A cryogenic setup is developed for measurements of the SNSPD performance in the broad spectral range of 400 – 1700 nm. The details of the measurement setup and the extraction of the detector properties from the data are described in section 6.1.

The performance of SNSPD in dependence on the material composition of the NbN film is investigated on a series of devices structured from the same films that were characterized in chapter 3. The impact of the material properties and the measurement temperature on the spectral detection efficiency are analyzed in the framework of the hot-spot model. In the final chapter, SNSPD with spiral shape are studied. The layout aims to reduce the current crowding effect on the basis of the requirements formulated in chapter 5. The critical current of such structures and their spectral detection efficiency are discussed.

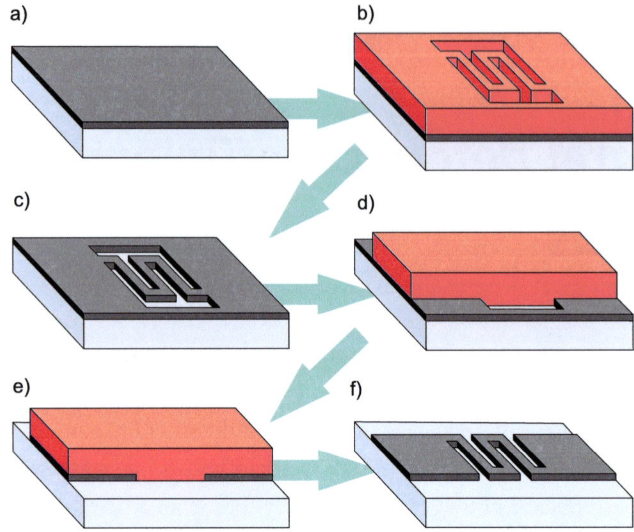

Figure 6.1: Schematic illustration of the patterning process of the SNSPD devices: a) The NbN thin film (gray) is deposited on a sapphire substrate (blue). A 95 nm thick resist layer (red) is applied by spin coating. The negative image of the central detector structure is exposed with the electron beam and removed during development (b)). The image is then transferred by reactive ion etching into the NbN film and the remaining resist stripped (c)). A second layer of resist is structured by photolithography in a positive image of the contacts with an additional part to cover the central structure (d)). After another reactive ion etching step (e)), the resist is removed (f)) to reveal the final device.

6.1 Characterization of the performance of SNSPD

6.1.1 Patterning of detector structures

The NbN thin films are patterned into SNSPD devices by a two-step process that is shown schematically in Fig. 6.1. First, the central detector part, consisting of the nanowire detector and the adjacent areas, are defined by electron-beam lithography. The image is then transferred into the NbN film by reactive ion-etching. In a consecutive photolithography, the coplanar contact pads of the devices are added around the detector. The same reactive ion-etching process is used for this step, where the central area that was structured before is protected by the resist.

Electron-beam lithography

In preparation of the first step, alignment marks are structured on the NbN film by a standard electron-beam lithography and lift-off process. They consist of a 40 nm Au film with a 10 nm thick Nb buffer layer to increase the adhesion, both sputtered *in-situ* on the substrate. The Au marks are used for high precision positioning in the following electron-beam lithography step, as well as for the alignment of the photo-lithography step later on. As resist for the electron-beam exposure, the film is spin-coated by a 95 nm thick polymethyl methacrylate (PMMA) layer, diluted with a ratio 2:1. In a write field of $100 \times 100\,\mu m$ around the central detector structure, the resisit is exposed in a negative image of the detector structure. Detailed exposure parameters are given in table 6.1. The exact exposure dose was determined for each film separately by a dose test directly prior to the exposure and was about 75 $\mu C/cm^2$. After development, the area around the structure is exposed once with the 20 μm aperture and 5 kV high voltage at a magnification of 5000 at scan speed 4. This post-exposure step enhances the stability of the resist and improves the edge quality.

Table 6.1: Electron beam lithography parameters for the center structures of the SNSPD devices.

Parameter	Value
Resist	PMMA 950 3% 2:1
Spin coating	60 s at 8500 rpm
Out-baking	5 min at 175°C
High voltage	10 kV
Aperture	10 μm
Writefield	$100 \times 100\,\mu m$
Dose	≈ 75 $\mu C/cm^2$
Developer	AR 600.56
Development time	30 s

Reactive ion-etching

The following etching procedure is crucial for the quality of the detector. The pattern transfer from the electron-beam resist to the superconductor film defines the edge roughness of the nanowires. Also, a detrimental effect of the etching procedure on the surface and edges that reduce the superconducting properties of those areas has to be minimized. A pattern transfer with dimensions below 50 nm to NbN thin films can be achieved with high quality by reactive ion etching in SF_6 plasma [150].

The process was first established for plane NbN and Nb films and then adapted for the special case of nanowire meander structures. In this case, the gaps between the nanowires are in the order of the resist height or smaller, so the aspect ratio influences on the etching rate. The optimal dose range for the electron-beam exposure was determined beforehand and four values inside this range selected for variation. After the post-exposure, the samples were then etched under different conditions and their resistive and superconducting parameters were measured.

Samples with obvious defects or connections between lines are sorted out after inspection by SEM. A more sensitive measure for the quality of the etching process is the room temperature resistance R of the sample. If the gap is etched incomplete, short-cuts remain that reduce the resistance of the structure significantly. Also, an over- or under-etched sample will have a final

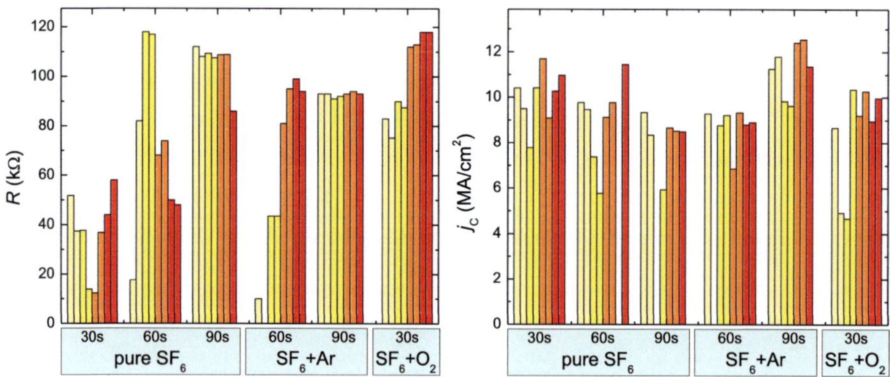

Figure 6.2: Room temperature resistance R (left) and critical current density j_C at 4.2 K (right) of meander test structures etched at various selected process parameters. Each bar represents one sample, where the bar color indicates the dose of the electron-beam exposure from low (light yellow) to high (red).

nanowire width that differs from the original resist width and thus shows a different resistance. Only process parameters where short-cuts can be excluded are considered further. As a second measure of quality, the critical current of the samples is considered. This parameter is more sensitive to the etching procedure and reveals the homogeneity and roughness of the line edges. Figure 6.2 shows the results for a selection of different process parameters. The first three sample series were etched with pure SF_6 with a flow of 20 sccm. Although already 30 s would suffice to remove the complete NbN film on a film without structured resist on top, for the case of the meander structures only after 90 s of etching the last remaining short-cuts are reliably removed. For the next two series, Argon is added to the SF_6 to provide an additional physical etching component. This should provide an increased anisotropy of the process. In this case, the gas mixture is 30 sccm SF_6 to 10 sccm Ar. Again, the etching time has to be increased from 60 s to 90 s for the meander samples to exclude short-cuts. In a final series the Ar was replaced by oxygen, where already after 30 s the resistance values were close to the nominal values. In each series the dose was varied in four steps indicated in Fig. 6.2 by the color of the bars. In the case of the O_2 process, the samples with low dose (light yellow) showed a lower R than the samples with high dose (red). For both the pure SF_6 process as well as the SF_6/Ar process, a well-reproducible R is achieved after 90 s. However, the latter also provides the highest critical current values of all series. This process is therefore adopted for the patterning of the SNSPD samples. At a plasma power of 100 W, and a process pressure of 90 mTorr, the etching rate of the NbN is determined to 0.21 nm/s. After stripping of the resist by acetone, the structures are investigated by SEM to measure the actual nanowire width and identify possible constrictions or defects. Figure 6.3

Figure 6.3: Schematic of the impedance matched coplanar layout, where black is the NbN film and white the sapphire substrate. The center conductor leads the RF signal to the readout (arrow). The detector device is placed at the other end of the conductor after it was tapered down. The zoom-in to the rigth shows an SEM image of an SNSPD. Here, light parts are NbN and dark parts the substrate.

shows an SEM image (right side) of a standard meander structure. The bright parts are the NbN surface, while darker parts are the bare substrate surface.

Contacts layout

In the last step, the contact pads are added by photolithography with standard parameters and reactive ion etching, where the central part that was already structured is protected by resist (Fig. 6.3, left). The layout of the contacts is designed as a coplanar waveguide, which consists of a center conductor framed by a ground plane. The width of the center conductor as well as the distance to the ground are engineered to yield an RF impedance of $50\,\Omega$ for the given dielectric substrate. The left side of the final sample, as shown in Fig. 6.3, will be mounted face-to-face to a circuit board with the same conductor width and gap dimensions, so that after connection the RF reflections at this point are minimal. On the sample side, the impedance is kept constant as long as possible by tapering both condutor and gap width continuously. The unavoidable impedance step from the $50\,\Omega$ of the readout electronics and the sample impedance occurs shortly before the meander begins. The coplanar waveguide design was developed and optimized by M. Hofherr and Dr. S. Wünsch of the Institut für Mikro- und Nanoelektronische Systeme at the KIT. The complete sample with the coplanar contacts has dimensions of $3\,\text{mm} \times 3\,\text{mm}$. After lithography, the samples are diced by a wafer saw into this size to be mounted into the measurement setup.

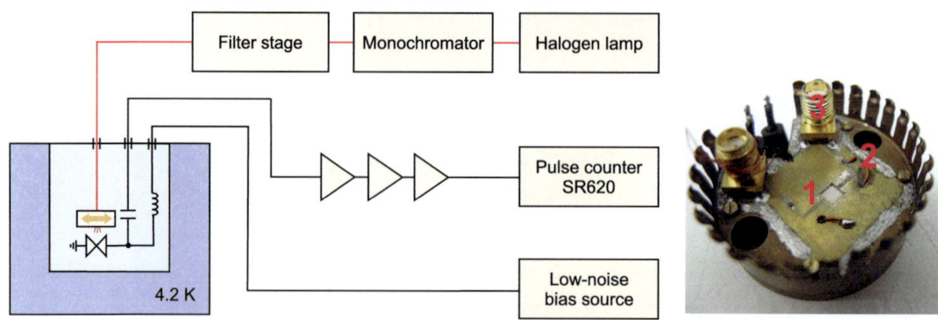

Figure 6.4: Left: Experimental setup scheme for the measurement of SNSPD in a spectral range from 400 – 1700 nm. The optical fiber (red) feeds the light from the monochromator into the cryogenic part and is mounted on a movable stage above the sample surface. The RF and DC paths are de-coupled at the sample stage and led out separately. The voltage pulses of the detector are amplified at room temperature before they are recorded by a pulse counter. Right: Photograph of the sample holder. (1) marks the grove for the sample that can be bonded to the readout line on the circuit board to the upper right, (2) is the bias tee and (3) the SMA connector for the readout cable. To the bottom left of the detector grove is an optional second readout channel.

6.1.2 Fiber-based experimental setup

The cryogenic setup for the measurement of SNSPD properties is based on a very similar setup that was developed in [151]. Its main body is a long vacuum-tight metal tube that can be inserted into a liquid Helium Dewar. The cooling power on the sample holder inside the tube can be varied continuously by introduction of contact gas. A heater element and temperature sensor on the sample holder allow the temperature to be controlled in the range between 4.2 and 20 K.

The light from a broad-band halogen lamp is passed through a monochromator to select the desired wavelength between 400 and 2200 nm. An optional stage allows the insertion of a wave-length filter before the light is coupled by a collimator into a graded index multimode fiber, which has a high transmission in the full output spectrum of the monochromator. The fiber is passed into the tube by a vacuum feed-through to the sample and ends a few mm above the sample surface. The position of the fiber end can be readjusted at low temperatures to compensate cool-down drift.

The samples are manually wire-bonded with indium to a circuit board where the high frequency path is decoupled from the bias path. An additional cold stage pi-filter reduces external interferences. The samples are biased by a battery-powered low-noise voltage source. The high frequency output signal is led out of the tube by stainless steel rigid coaxial cables. A 6 dB attenuator limits the back-reflections from the amplifier chain. The signal is amplified at room temperature by a set of three 2.5 GHz amplifiers and then sent to a pulse counter with 300 MHz internal bandwidth.

Fig. 6.5 shows a typical output pulse of an SNSPD count event in high resolution. The measurement was performed at the experimental setup described above, but the pulse counter was

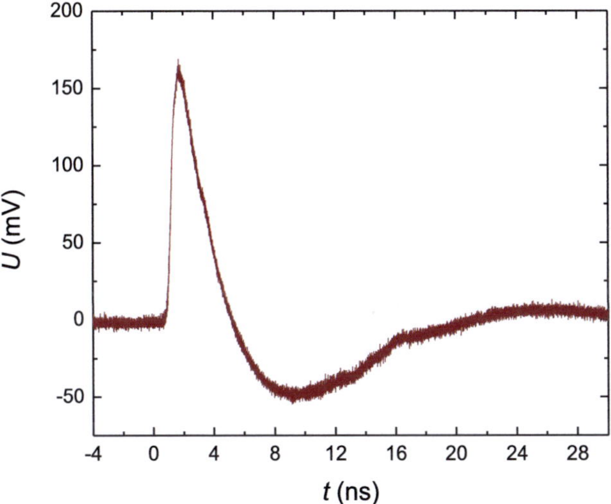

Figure 6.5: Typical output voltage pulse of an SNSPD count event. The pulse was measured with a self-triggered 20 GHz sampling oscilloscope after the amplifier chain. The FWHM of the pulses are in the ns range, depending on the sample inductance.

replaced by an RF splitter and a 20 GHz sampling scope, where the second path of the splitter was used for self-triggering. The pulse thus directly reflects the output of the detector with the bandwidth limitation of the amplifier chain. Typical SNSPD pulses have a full-width half maximum (FWHM) in the nanosecond range, depending mainly on the kinetic inductance of the nanowire. This inductance and the capacitance of the bias tee form an LC-resonator, which is responsible for the oscillatory shape of the pulse after surpassing the maximum amplitude. The pulses are well-resolvable above the system noise level approximately for bias currents above $5\,\mu A$. However, the exact pulse shape is dependent on various sample and measurement conditions and will in the final setup be distorted by the lower internal bandwidth of the pulse counter. It is therefore advisable to control the resolvability of the pulses for each sample separately by observation of the count rate for different trigger levels at the pulse counter, as described in the next section.

6.1.3 Measurement of detection efficiency

The exact definition of this parameter varies in literature, depending on how much of the optical coupling efficiency is already accounted for. In this work the detection efficiency (DE) is the ratio of the count rate of voltage pulses caused by photon events to the rate of photons n_{ph} incident on the detector area. The rate of voltage pulses not caused by the photon events is the so-called dark count rate *DCR*. It has to be evaluated separately and subtracted from the total count rate n. The detection efficiency is then given by

$$DE = \frac{n - DCR}{n_{ph}}.$$

(6.1)

All losses of coupling the light into the fiber, losses in the fiber itself as well as light that is not incident inside the detector area are not included in this definition and have to be accounted for by careful optical calibration.

On the other hand, the filling factor of the detector area, the polarization dependence as well as the absorptance of the NbN are part of the detection efficiency. Their influence and the connection to the intrinsic detection efficiency is discussed in section 6.1.4. Here, the evaluation of the count rate n, the dark count rate *DCR* and the calibration methods for the determination of the photon flux n_{ph} are discussed.

Count rate

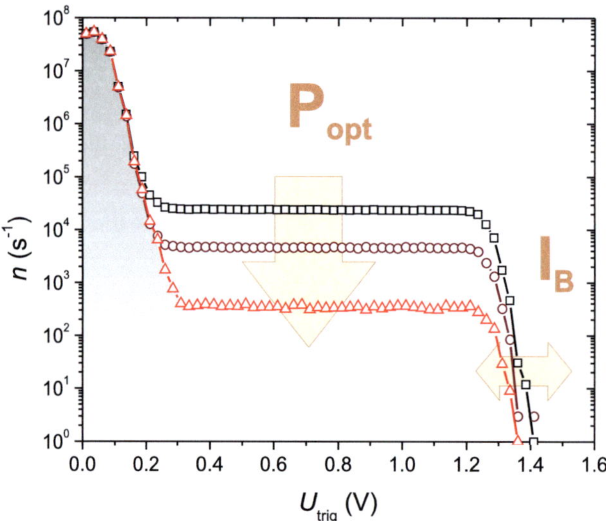

Figure 6.6: Count rate in dependence on the trigger level of the pulse counter. For the grey area, the trigger level is below the noise level. The count rate on the trigger plateau depends on the incident optical power P_{opt} (black: high power, brown: medium power, red: low power). The plateau ends when the trigger level exceeds the pulse height, which mainly depends on the bias current I_B (not shown).

Once a count event is triggered in the detector, a voltage pulse is generated and led to a pulse counter. The device counts all rising flanks exceeding a certain trigger level U_{trig} during a set time interval. To assure correct count rates, U_{trig} is defined for each detector individually. Figure 6.6 shows a typical dependence of the count rate on the set trigger level for different optical powers P_{opt} incident on the detector. The gray shaded area for very low U_{trig} is dominated by the voltage noise of the system. For higher U_{trig} a plateau emerges that reflects the rising flank of a count event and gives the correct count rates n. Since n scales with the incident optical power P_{opt}, the point of cross-over from the noise region to the plateau varies. Once U_{trig} exceeds the height of the voltage pulse, the count rate drops to zero with some statistical distribution of the actual peak heights, marking the end of the plateau. This value depends directly on the bias current I_B, which also influences on the count rate n itself.

U_{trig} thus needs to be low enough to be still on the plateau for the lowest considered I_B, but large enough to have insignificant noise contributions for the lowest expected P_{opt} at this bias current. For detector structures with high critical currents, this conditions can generally be met in a wide margin of U_{trig}. However, it becomes increasingly restrictive for devices with low critical currents, i.e. small voltage amplitudes. For the system described here, this was roughly the case for structures with $I_C^m < 10\,\mu\text{A}$, depending on the exact measurement conditions. Once evaluated, U_{trig} is kept constant for all measurements on the sample.

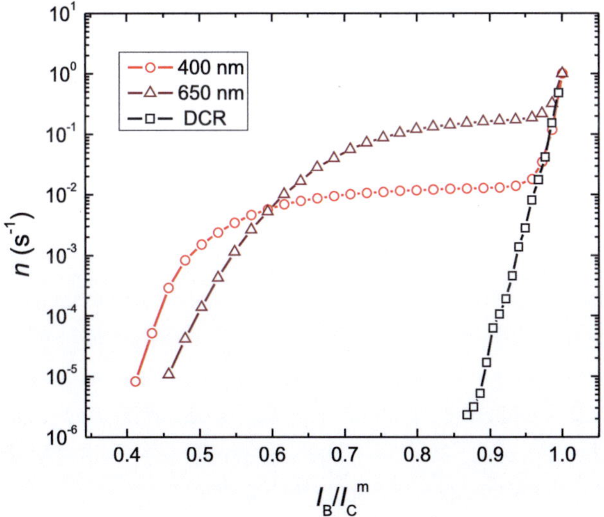

Figure 6.7: Count rate on reduced bias current of an SNSPD illuminated at 400 nm (red) and 650 nm (brown). The dark count rate (black squares) was measured with a blocked beam path and rises exponentially. Close to $I_B/I_C^m = 1$, it dominates the count rates and causes the upturn.

The dark count rate DCR is measured when the light is blocked. This is realized at the fiber vacuum feed-through where the fiber can be disconnected. The DCR rises exponentially as the bias current I_B approaches the critical current I_C^m of the detector. Close to I_C^m it thus dominates

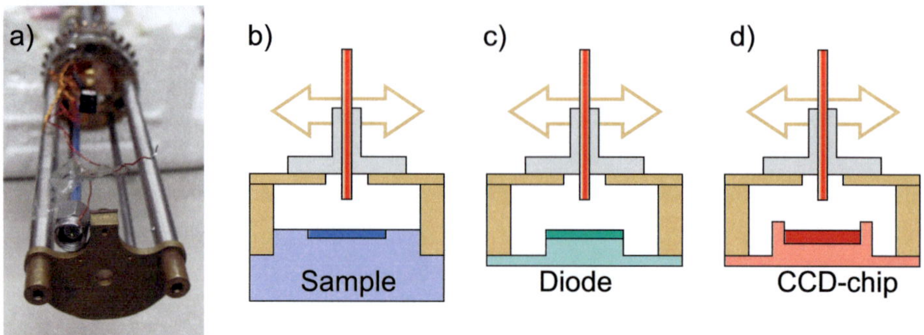

Figure 6.8: a) Photograph of the setup fiber end without an equipped modular stage. The fiber is located in the hole of the face plate and its position can be altered in two dimensions by the movable carriage on the other side. b) Schematic of the fiber stage and the modular sample holder. The detector sample (dark blue) is located opposite the fiber end in a certain adjustable distance. The whole blue block can be replaced by calibration blocks, where either photodiodes (c)) or a CCD chip (d)) are located in place of the sample.

the count rate under optical irradiation n and causes an upturn (see Fig. 6.7). The count rates n show an otherwise relatively flat dependence on I_B/I_C^m, which indicates that here each absorbed photon has enough energy to trigger a count event. This corresponds to the hot-spot detection mechanism described in section 2.2.1. Below a certain I_B/I_C^m value, the energy is not sufficient any more and the count rate drops. For less-energetic photons of a longer wavelength, this point is reached already at higher I_B/I_C^m (The different height of the count rates in Fig. 6.7 are due to different P_{opt} of the two wavelengths).

Photon flux

The rate of photons n_{ph} can be directly calculated from the optical power incident on the detector area $P_{det} = n_{ph}hc/\lambda$. To determine P_{det}, the power density distribution in the detector plane is determined and then integrated over the detector area A_{det}. This is done in two steps: First, the total optical power that is emitted by the fiber end P_{out} is determined by highly sensitive area integrating photodiodes. Second, the relative intensity profile is measured with a CCD camera and calibrated with P_{out} to give the power density distribution.

For these calibration purposes, the cryogenic setup is equipped with a modular end stage. For normal measurements, the sample holder is placed at the end stage which contains the circuit board and RF and DC connectors (see Fig. 6.4 (right)). Opposite the sample holder, the multi-mode fiber end is mounted on a movable carriage. The fiber is cleaved with 7° angle to minimize reflections. The distance of the fiber end from the sample surface h can be changed when the setup is open and is then fixed for the measurement. The x,y-direction (parallel to the detector plane) can also be changed at cryogenic temperatures with a travel range of ≈ 1 mm, to correct cool-down drift. Figure 6.8 a) shows a photograph of the end of the setup with the fiber carriage. On this end stage, the the sample holder can be mounted as schematically shown in b), so that the fiber position relative to the sample position remains variable. For calibration purposes, the sample holder can be replaced by photodiode modular stages (c)), where the diodes are placed at the identical position as the sample in the sample holder. Several types of photodiodes cover the wavelength region from 400 – 1700 nm and are used to determine the output power at the fiber end $P_{out}(\lambda)$ relative to the input power $P_{in}(\lambda)$ of the system: The visible range (450 – 750 nm) is covered by a Si-diode, the infrared range (750 – 1700 nm) by a Ge-diode and the small UV part (400 – 450 nm) can be complemented by a GaP-diode. All diodes have areas that far exceed the light spot dimensions on the petector plane for fiber distances up to $h = 5$ mm. Since the power spectrum of the monochromator depends strongly on the selected grating and is further not reproduced well after a filter change, the input power spectrum $P_{in}(\lambda)$ is determined directly before each measurement.

An additional calibration end stage is similarly equipped with a CCD chip in the detector plane (Fig. 6.8 d)). Since the chip is less sensitive than the photodiodes, the intensity profile is taken at much higher incident optical powers; however, the shape of the profile scales with the incident optical power. The two-dimensional integral of the intensity profile is equated with the P_{out} value determined by the photodiode to acquire the power density distribution in the detector plane (for details refer to [152]).

A typical beam intensity profile measured by the CCD camera is shown as inset in Fig. 6.9. The cross-sections of the power density distribution through the center of the radially symmetric beam profile (dashed gray line) is displayed for various distances h between fiber and detector plane. The beam radius is already for relatively small h of ≈ 1 mm much larger than any detector area considered. Since the multimode fiber has a core diameter of 105 μm, it can not be treated as a point source. Instead, its index step function is convoluted with a Gaussian function to give the expected beam profile. In contrast to a pure Gaussian distribution, this profile has a relatively broad, almost flat region in its center which leads to a homogeneous illumination of a detector placed at $x = 0$. Indeed, the measured profiles in Fig. 6.9 match this expected shape very well. To ensure the detector is actually at the desired position even at measurement conditions, the detector count rate is determined in dependence on the x,y position of the fiber. This directly reflects the known intensity profile from the CCD camera and the fiber can be positioned correctly. The intensity profile itself has a slight wavelength dependence. The general shape is very similar, but the spot size, defined at 1/e of the maximum, is reduced for larger wavelengths. From 450 to 950 nm, it is decreases by about 7%. The respective change in the intensity distribution at sample position leads to a slight correction factor in the calibration.

It has to be taken into account that the light from the monochromator is polarized due to the diffraction grating. Meander-shaped SNSPDs have been shown to have polarization-dependent detection efficiency [19]. This is due to their geometry, where the long parallel wires act as a wire-grid polarizer. With other layouts like partly perpendicular meanders or spirals, the po-

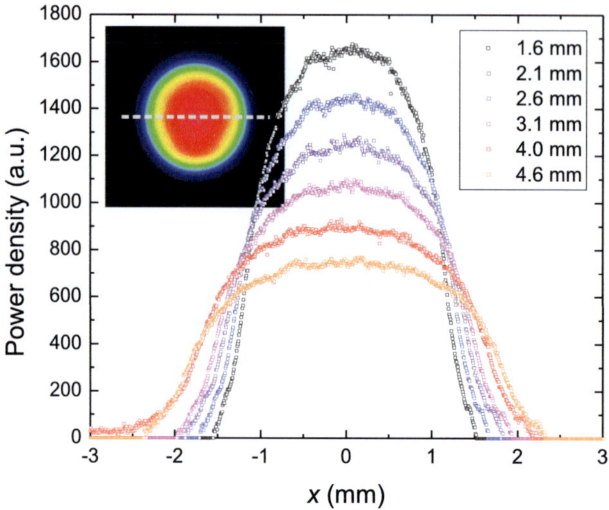

Figure 6.9: Cross-section of the power density distribution for varying distances between fiber end and detector plane h as indicated in the legend. $x = 0$ defines the lateral detector position. The inset shows an intensity profile measured with a CCD camera at $\lambda = 650\,\mathrm{nm}$, where high intensity is red and low intensity is blue/ black. The cross-section of the radially symmetric density distribution is evaluated along the dashed grey line.

larization dependence can be minimized [153]. Additionally, it has been found that even the intrinsic detection efficiency of the device is higher for light polarized parallel to the wire [154]. Since the multimode fiber used in this setup is not polarization maintaining, it is not possible to control the polarization state of the light at the sample position. However, measurements of the grade of polarization $(I_p - I_s)/(I_p + I_s)$ reveal that it reduces with fiber length. For the SNSPD characterization, the fiber length between monochromator and setup is increased to $l = 2\,\mathrm{m}$, for which the grade of polarization at detector position becomes lower than 2%.

Since the beam spot, especially for large h, is much larger than the detector area, most of the optical power is absorbed by the surface of the sample holder and introduced heat into the system. If P_{in} is too large, so this exceeds the cooling power of the system, this will influence on the measurement by reducing the critical current I_C^m of the sample. This defines an upper limit of P_{in} for the experiment. For optical power below this limit, the count rate n of the detector scales linearly as is shown in Fig. 6.10. For typical measurements, powers of roughly $P_{in} = 100\,\mathrm{nW}$ are used, which for a standard detector design corresponds to $P_{det} \approx 5\,\mathrm{pW}$. This amount of light is in the regime of single-photon excitation, demonstrating the single-photon sensitivity of the devices.

The spectral dependence of the number of photons incident on the detector plane after application of the complete calibration and corrections is shown as solid points in Fig. 6.11 (left). The

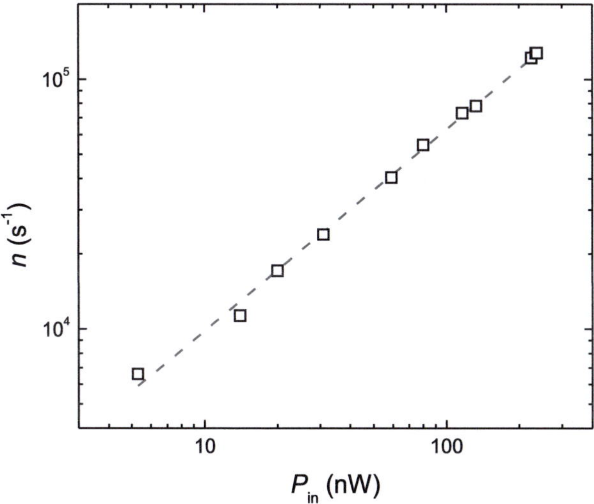

Figure 6.10: Proportionality of SNSPD count rate to optical power P_{in} that is coupled into the measurement system. The dynamic range of the detector spans more than two orders of magnitude.

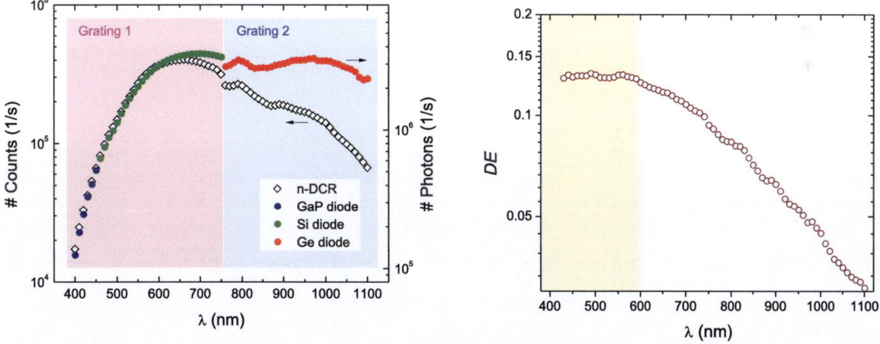

Figure 6.11: Left: Spectral dependence of a typical SNSPD count rate $n - DCR$ (open symbols, left axis) in comparison with the photon number incident on the detector area in the same time interval (solid symbols, right axis). The diode type used for the calibration of a wavelength range is given in the respective color in the legend. At $\lambda = 750$ nm, the grating of the monochromator and the band pass filter are changed, which causes the offset in the photon number. Measurements are shown here only up to 1100 nm, but can be extended after another filter change up to 1700 nm. Right: Dividing $n - DCR$ by the photon flux yields the spectral detection efficiency. Shown here is the DE of a 4 nm thick SNSPD biased at $0.95I_C^m$. The yellow shaded region marks the wavelengths with hot-spot detection mechanism, where DE is independent of the photon energy. For larger wavelengths, DE decreases steadily.

color of the points corresponds to the photodiode that was used for the calibration of the respective wavelength range. The number of counts n of a typical SNSPD measurement are shown as open points, already corrected for the dark count rate DCR. With the known photon flux and its spectral dependence, the detection efficiency can now be calculated from the count rates for each specific wavelength and bias current (Fig. 6.11 (right)). The accuracy of the current state of the described system, taking all influences discussed above into account, allows the calculation of DE with an error below 4%.

6.1.4 Spectral detection efficiency and intrinsic detection efficiency

Figure 6.11 (rigth) shows a typical spectral detection efficiency for a standard SNSPD made from a 4 nm thick NbN film, biased at $I_B = 0.95 I_C^m$. The DE reaches its maximum value in the low-wavelength optical region, where the dependence is relatively flat and independent of the photon energy. This corresponds to the hot-spot detection mechanism described in section 2.2.1. Beyond the cut-off wavelength λ_C, the detection efficiency drops significantly when the detection mechanism changes to the fluctuation-assisted photon detection as was discussed in section 2.2.2. The exact position of λ_C depends on the cross-section of the nanowire, the applied bias current and the material properties of the superconductor. The latter will be discussed in detail in section 6.2.

Detection efficiency spectra measured at different bias currents I_B are shown in Fig. 6.12. All other measurement conditions were kept constant. For a higher applied bias current, the hot-spot region extends further into the infrared range. Above λ_C, this leads to a higher total DE for the higher bias currents. However, the dark count rate rises also with the bias current and limits the range of feasible biasing conditions. The cross-section of the nanowire displays an analog influence on the detection efficiency spectra: For lower film thickness and narrower nanowire width, the hot-spot plateau extends further into the infrared range.

Intrinsic detection efficiency

The detection efficiency DE consists of two factors: The absorptance ABS of the structured NbN film and the intrinsic detection efficiency IDE of the material itself

$$DE = ABS * IDE. \tag{6.2}$$

Since the physics of the detection mechanism is contained in IDE, it is necessary to remove the influence of the wavelength-dependent absorptance from the detection efficiency in order to determine the cut-off wavelength. One way to solve this problem which is pursued in the literature is to evaluate $ABS(\lambda)$ carefully and then calculate IDE. The absorptance of NbN thin films and its thickness dependence was discussed in section 3.5. However, it has been shown [19] that $ABS(\lambda)$ of structured SNSPDs differs from that of an unstructured thin film and is also strongly polarization dependent. This is not unexpected, since the detectors are structured

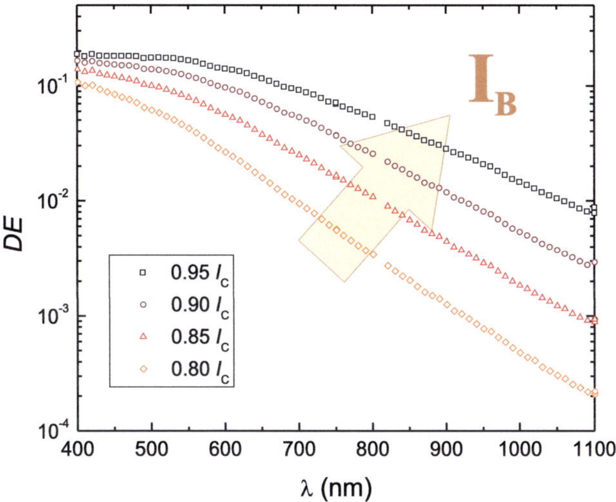

Figure 6.12: Spectral detection efficiency of an SNSPD for varying bias currents given relative to the critical current I_C. With increasing bias ratio, the spectrum extends farther into the larger wavelength range and thus exhibits larger DE values in the infrared range.

in sub-wavelength dimensions and the meander-lines give a clear directional preference. The absorptance of an unstructured film can thus only give an indication of the relative behavior, but is not exact enough for the calculation of IDE.

Another method to remove ABS is utilizing the fact that it is independent of the bias current I_B. First, the detection efficiency DE_1 is measured at a high bias current, e.g. at $I_{B1} = 0.95 I_C^m$. A second detection efficiency DE_2, measured at a lower $I_{B2} < I_{B1}$ can then be normalized to DE_1 and ABS is canceled out:

$$\frac{DE_2}{DE_1} = \frac{ABS\,IDE_2}{ABS\,IDE_1}.$$

(6.3)

In the range up to the cut-off wavelength of I_{B1}, IDE_1 is constant and thus $DE_2/DE_1 \propto IDE_2$ directly shows the wavelength dependence of the intrinsic detection efficiency for I_{B2} (see Fig. 6.13). Above λ_{C1}, IDE_1 starts to decrease, leading to an upturn in DE_2/DE_1.

Cut-off wavelength

The spectral dependence of the IDE can be formally described by

$$IDE \propto \frac{1}{1 + (\lambda/\lambda_0)^p},$$

(6.4)

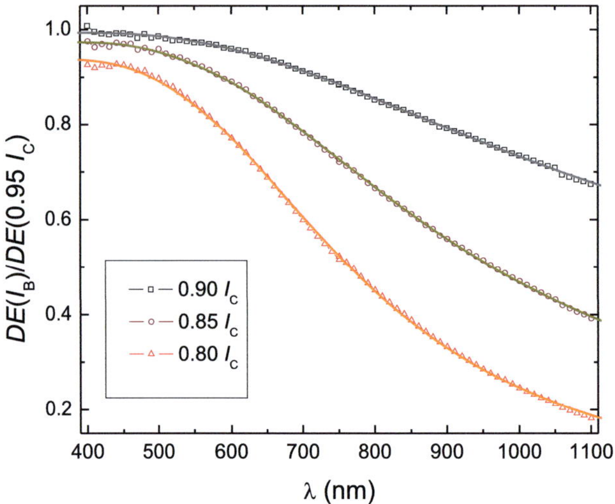

Figure 6.13: Normalized detection efficiency spectra for a 4 nm thick, 100 nm wide NbN SNSPD. The curves correspond to three different I_B as indicated in the legend and are all normalized to $DE(0.95I_{C,m})$. The solid lines are fits of eq. 6.5 to the measurements.

where λ_0 is the cut-off wavelength and p describes the power-law dependence of IDE in the infrared range. This definition of the cut-off is slightly different than the ideal abrupt cut-off λ_C predicted by the hot-spot model, as it marks the wavelength for which the IDE has been reduced to 50% of its maximum value. However, it provides a very robust definition of the cut-off wavelength, and the relative behaviour of λ_c and λ_0 should be identical.

To determine the cut-off wavelengths from the normalized detection efficiency spectra, they are fitted by the expression

$$\frac{DE_2}{DE_1} = \frac{1+(\lambda/\lambda_{0,1})^{p_1}}{1+(\lambda/\lambda_{0,2})^{p_2}}, \tag{6.5}$$

where the parameter sets 1,2 for the two bias currents $I_{B,1}$ and $I_{B,2}$ are the fitting parameters (solid lines in Fig. 6.13). In this way it is possible to extract one cut-off wavelength value $\lambda_0(I_B)$ for each applied bias current.

With the presented experimental setup and methods, the dark count rate, detection efficiency and the cut-off wavelength for detector samples can be evaluated. The following sections describe several series of measurements that have been taken on SNSPD samples and results are discussed.

6.2 Influence of the NbN stoichiometry on the performance of SNSPD

The variation of the NbN thin film parameters with the change of material composition revealed in chapter 3 already give an indication that an influence on the detector performance of SNSPDs can be expected. However, to predict the outcome remains difficult due to the multitude and partly contrary dependences of parameters. To investigate which parameter is dominant in the processes that lead to a detection event, a series of 4 nm thick films deposited at four distinct sputter currents I_{SP} were selected to be structured into SNSPD to directly measure the detector performance. The results discussed in the following were published in [85].

As a reference sample, a film deposited at I_{SP} = 145 mA was selected. For this current the optimal critical temperature is reached (compare Fig. 3.6), which is the standard deposition condition for most NbN films found in the literature. The next sample was taken at the maximum of the critical current density (compare Fig. 3.10), which for the 4 nm thick series was at I_{SP} = 175 mA. The two remaining films are deposited at the extreme conditions I_{SP} = 100 mA for large excess nitrogen and I_{SP} = 190 mA with nitrogen deficiency. For higher or lower sputter currents, T_C becomes too small for feasible SNSPD application.

According to the investigations presented in chapter 3, the material properties are quite sensitive to the sputter conditions of the NbN deposition and the the film thickness. To make a reliable comparison possible, the detectors are structured directly from the same films that have been investigated there, so that the known film parameters can be used in the calculations. The thickness of the films was measured by profilometry to be 4.3±0.4 nm. The patterned films showed a decrease of the critical temperature T_C of about 1 K as compared to the unstructured films due to proximity effects of the narrow lines. The sheet resistance R_S, critical temperature T_C and the critical and hysteretic current densities are listed in table 6.2. Further listed in the table are film parameters extracted by measurements described in detail in chapter 3: The electronic diffusion coefficient D, the coherence length ξ and the energy gap 2Δ at 4.2 K.

From each film, nine nanowire detector samples were patterned with a nanowire width of 100 nm, a gap between the nanowires of 60 nm and a detector area of 4x4.2 μm^2. By inspection with SEM, structures with defects or constrictions were sorted out and the line-width w of the wire was confirmed to be within 5 nm of the nominal value.

6.2.1 De-pairing critical current

According to eq. 2.6, the spectral bandwidth of an SNSPD depends on the ratio I_C^m/I_C^d. The critical current of the detector in the experimental setup, I_C^m, is defined by the clear discontinuity of the voltage when increasing the current from the superconducting state. It is in principle possible in a nanowire to reach the theoretical limit of the de-pairing critical current I_C^d, which is the current at which the Cooper pair kinetic energy exceeds the binding energy. However, for

Table 6.2: Average values of SNSPD samples made from NbN thin films with varying chemical composition. The width of the nanowire is $w = 100$ nm.

I_{SP}	d	R_S	T_C	$j_C^m(4.2\,\text{K})$	$j_H(4.2\,\text{K})$	D	ξ	2Δ	$\kappa_{C,0}$
mA	nm	Ω	K	MA/cm^2	MA/cm^2	cm^2/s	nm	meV	nm
100	4.7	510	10.5	3.3	0.85	0.47	4.3	4.43	237
145	4.2	430	12.5	4.6	1.07	0.53	4.4	3.68	224
175	3.9	291	11.8	6.0	1.33	0.61	4.8	3.80	153
190	4.4	254	10.0	4.1	0.81	0.70	5.5	3.38	180

standard measurement conditions of an SNSPD, I_C^m typically stays well below I_C^d. Reasons for this reduction were investigated and discussed in chapter 5.

To evaluate the respective ratio, I_C^m is directly measured on the sample in the experimental setup, while I_C^d is calculated from the material parameters given in table 6.2. Figure 6.14, left, shows $I_C^m(T)$ of one of the detectors (open symbols) in the temperature scale of the de-pairing current $(1 - T/T_C)^{3/2}$ in comparison with the theoretical dependence (solid line). $I_C^m(T)$ coincides with the de-pairing limit for temperatures close to T_C well, but below $\approx 0.85 T_C$ starts to deviate and stays below I_C^d similar to what has been reported in [28]. All samples of the different films show the same principal behavior, only the quantity of reduction of I_C^m in the low temperature region varies. It is thus sufficient to compare the values at 4.2 K, where the detectors are operated and the difference between I_C^m and I_C^d is most significant. Figure 6.14 shows the ratio of the experimental critical current I_C^m at 4.2 K normalized to the calculated de-pairing critical current I_C^d for the same temperature as a function of the film deposition current I_{SP}. Within samples from the same film there is a certain deviation due to individual fabrication uncertainties. There is however a clear general trend to higher I_C^m/I_C^d values for the films deposited at higher I_{SP}, i.e. with reduced nitrogen content. For the film with excess nitrogen ($I_{SP} = 100$ mA) only about 30% of the de-pairing critical current can be reached, whereas the ratio can be almost doubled to above 55% for $I_{SP} = 190$ mA. The results correlate well with similar measurements on NbN SNSPD with standard composition, where the measured critical current amounted to approximately 43% of I_C^d, shown as red point in Fig. 6.14 (right) [28].

A possible explanation for this dependence can be given by considering the critical current reduction induced by the sharp 180° turns of the meander line, as discussed in section 5.2. Although all detectors have the same absolute layout, the effect reduces with decreasing ratio of nanowire width to coherence length w/ξ. The width $w = 100$ nm is constant for all samples, but $\xi(0)$ increases for films with reduced nitrogen content (compare Fig. 3.12). The film with minimum nitrogen content ($I_{SP} = 190$ mA) has the highest value of $\xi(0)$ and the lowest T_C, giving it the highest $\xi(4.2\,\text{K})$. Consequently the ratio $w/\xi(4.2\,\text{K})$ is smallest and the current crowding effect weaker, leading to the highest I_C^m/I_C^d values, as confirmed by the experiment.

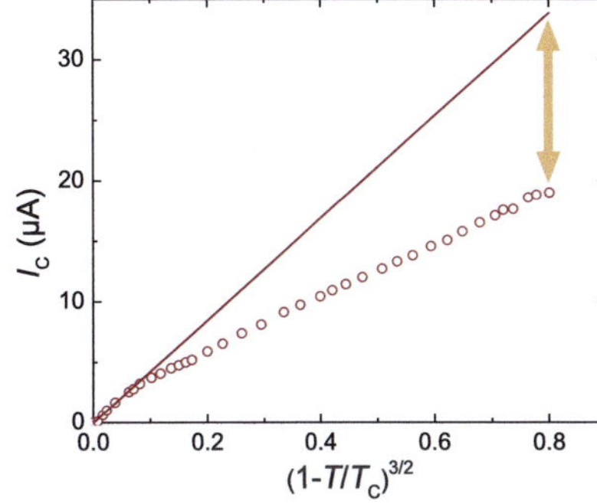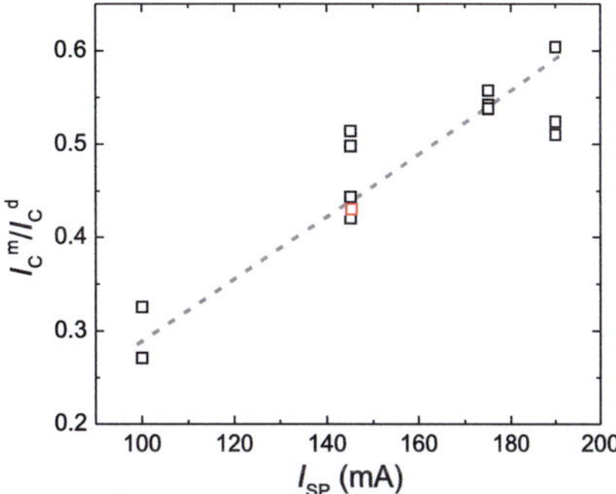

Figure 6.14: Left: SNSPD critical current I_C^m dependence on temperature, scaled to the de-pairing current temperature dependence. For $T \to T_C$, the data follow the de-pairing current dependence (solid line) and start to deviate at lower temperatures. For 4.2 K (last data point to the right), the ratio of I_C^m to the calculated value of I_C^d is evaluated for each detector. The results are shown in the right graph for the four different sputter currents I_{SP} at which the films were deposited. The trend to higher ratios for higher I_{SP} is illustrated by the dashed line. The red data point is a comparable measurement on a SNSPD with the same film thickness and NbN standard composition, given in reference [28].

6.2.2 Dark count rate

Count events of SNSPD can still occur, even if there are no optical or infrared photons incident on the detector. This event is termed a dark count event and the respective rate (dark count rate, *DCR*) is the analog to noise in a power detector. It sets the limit for the minimum count rate, and thus the minimum average radiation power that can be resolved.

The origin of dark count events can be black-body radiation from surrounding surfaces, fluctuations of temperature and bias current, or statistic events like vortex penetration and movement in the superconducting nanowire. These are the same fluctuations that give rise to the fluctuation-assisted photon detection in the infrared range as discussed in section 2.2.2. Contributions from black-body radiation are often encountered when the optical fiber is very close to or in contact with the sample surface, and especially emerge when cooling further below 4 K [155, 156]. They have the same dependence on I_B as regular counts from white light irradiation with low power. We do not observe such an influence on the *DCR* down to the lowest count rates that are recorded and thus neglect this portion. The main contribution, especially for I_B close to I_C^m, is the thermal excitation of magnetic vortices across the edge barrier of the nanowire and the consecutive dynamics caused by the Lorentz force of the bias current acting on the vortices. The probability of a thermal excitation of vortices is proportional to $I_B \exp(-U(I_B)/k_B T)$ with the bias-dependent potential barrier $U(I_B)$ given in eq. 2.7.

Being thermally activated, the *DCR* depends on the thermal coupling of the nanowire to its environment. There are two principal cooling channels for any thermal energy generated in the wire: The first is the phonon transport to the surrounding atmosphere. In most experimental situations the detector is either in vacuum (e.g. in a bath cryostat) or in a controlled Helium gas atmosphere. Even if the pressure of this atmosphere is relatively high, i.e. close to ambient pressure, the cooling efficiency is relatively low compared to the phonon transport in solid materials. Thus it only needs to be considered in the case of a solid layer in top of the wire, for example optical cavities or anti-reflection coatings.

The second channel is the phonon coupling to the substrate material and further to the sample holder, which is typically made of metal with high thermal conductivity. This transport is mainly defined by the thermal conductivity of the substrate material and the interface between superconductor and substrate and between substrate and sample holder. If the contact has a low thermal resistance, fluctuations that occur are more strongly coupled to the large thermal bath of the sample holder. Detectors mounted on the holder with silver paste, which has a four orders of magnitude higher thermal conductivity than silicon grease, show in direct comparison a more than two orders of magnitude reduced DCR [157].

The time constant of phonon escape to the substrate and its dependence on the material was discussed in chapter 4. This contribution is additionally influenced by the geometry of the detector, in particular by the distance between the nanowires. If they are far apart, each nanowire would have a half space as available phonon angles for their out-diffusion into the substrate. When spaced closely together, the shallow angles overlap and the effective phonon diffusion is reduced.

Based on the considerations above, the NbN film properties can influence in two ways on the dark count rate of an SNSPD: First, the variation in thermal boundary between film and substrate can lead to different cooling efficiencies (The thermal conductivity of the material can be neglected because of the thin film thickness in case of an SNSPD). Second, the material properties influence on the potential barrier for vortex penetration into the nanowire.

In section 4.3.2, it was found that the energy relaxation time is only slightly material dependent and indicates a slightly improved thermal transport for the films with increased nitrogen content. As an additional quantification for the change in the thermal boundary interface, we compare the hysteresis current density j_H of the samples. The hysteresis current (also called re-trapping current) is the current at which Joule heating and cooling power in the nanowire are equal and thus gives a measure of the phonon escape efficiency [78, 158]. For all samples, $j_H(4.2\,\mathrm{K})$ is about $1\,\mathrm{MA/cm^2}$ and does not change significantly with the deposition current I_{SP} of the film. The slight variation that can be discerned follows mainly the dependence of the critical current density on T_C and shows a maximum at $175\,\mathrm{mA}$. From both observations, we can conclude that the thermal coupling to the substrate is not influenced significantly by the chemical composition, and we expect the main contribution to the DCR to be from vortex penetration.

Figure 6.15 shows four representative dependencies of the DCR on reduced bias current for films with different I_{SP}. The values of the dark count rates close to I_C^m vary slightly from detector to detector made from the same NbN film due to almost unavoidable differences in the detector

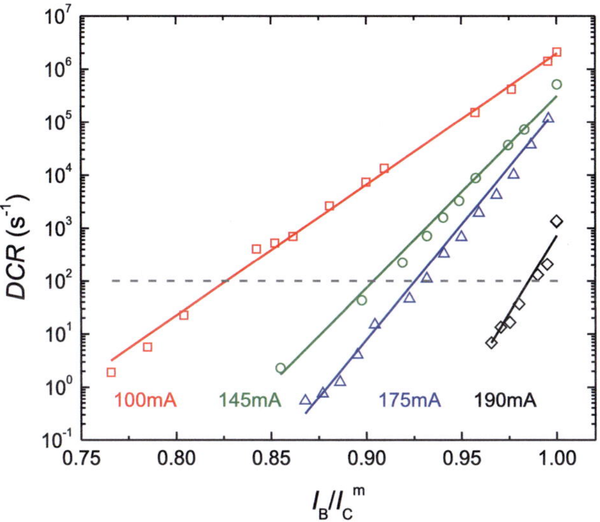

Figure 6.15: Dark count rate determined at 4.2 K on SNSPD structured from NbN films. The sputter currents at which the films were deposited are indicated in the graph. The solid lines are exponential fits to the data. The gray dashed line marks a DCR of $100\,\mathrm{s}^{-1}$, which is typical threshold of acceptable DCR.

mounting onto the sample stage as discussed above. In all cases, the detector was glued with conductive silver paste onto the brass sample holder and connected electrically with indium. The maximum DCR values are reduced almost by three orders of magnitude from the film deposited at 100 mA to the one deposited at 190 mA. Moreover, the DCR starts to have significant influence only at higher I_B for films with reduced nitrogen content: A DCR of 100 s^{-1} (dashed line) is reached at 0.83 I_B/I_C^m for the 100 mA-deposited film but only at 0.98 I_B/I_C^m for the 190 mA-deposited film. Exponential fits to the data (solid lines in Fig. 6.15) show a continuous increase of the slope for films with reduced nitrogen content, indicating an enhanced potential barrier for vortex penetration in those films.

The reduction of the DCR is a crucial factor in the improvement of SNSPD performance. There is a general trade-off in SNSPD between detection efficiency and DCR, which both rise with the bias currents applied to the detector. However, for a steep increase in the DCR it is still possible to come to the saturation region of the detection efficiency and achieve almost maximum values without allowing for too many dark count events. Detectors from films with reduced nitrogen content show a clear advantage in the range of bias currents with acceptable DCR.

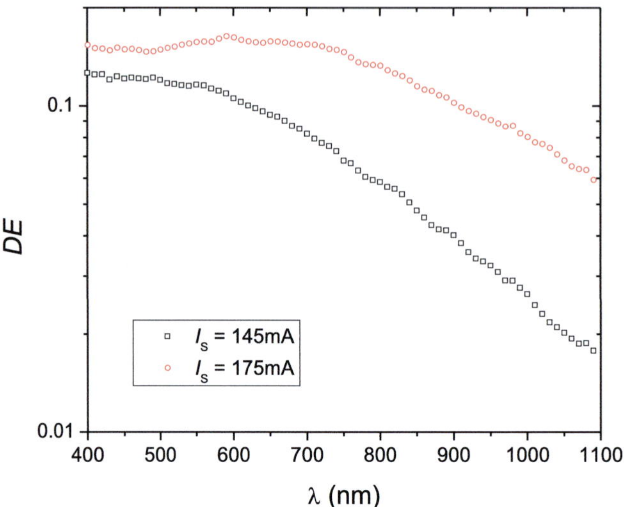

Figure 6.16: Detection efficiency spectra for two SNSPD measured at the same conditions $T = 4.2\,\mathrm{K}$ and $I_B = 0.9I_C^m$. One detector was made from a NbN film with standard composition (black), while the other detector was made from a film with reduced nitrogen content (red).

6.2.3 Improvement of the detection efficiency in the infrared spectral range

For the evaluation of the spectral bandwidth, 2-4 detectors with parameters close to the mean values were selected for each film. The spectral detection efficiency was measured on each detector as described in section 6.1 for several I_B values in the range from $0.8I_C^m$ to $0.98I_C^m$. Figure 6.16 shows the absolute spectral DE measured at $I_B = 0.9I_C^m$ for a detector with standard NbN composition (black squares) and one made from a film with reduced nitrogen content (red circles). In the hot-spot regime, i.e. for wavelengths below $\approx 550\,\mathrm{nm}$, the first detector shows an DE of 13%, whereas the second one has in this range a slightly improved value of 16%. This increase is within the limits of the accuracy of the determination of DE, which is 4 percentage points. Thus for this range, no significant increase in DE can be claimed. In the infrared range, however, the DE of the detector made from the film with reduced nitrogen is clearly superior. This is due to the much larger cut-off wavelength which leads to a broader plateau in the optical region.

In section 3.5 it was shown that the absorptance of the NbN films depends on the stoichiometric composition of the films. In order to exclude this contribution and to compare only the influence of the material on the intrinsic detection efficiency spectrum $IDE(\lambda)$, the cut-off values λ_0 are extracted from the DE spectra by normalization and fitting of eq. 6.5 as described in section 6.1.4.

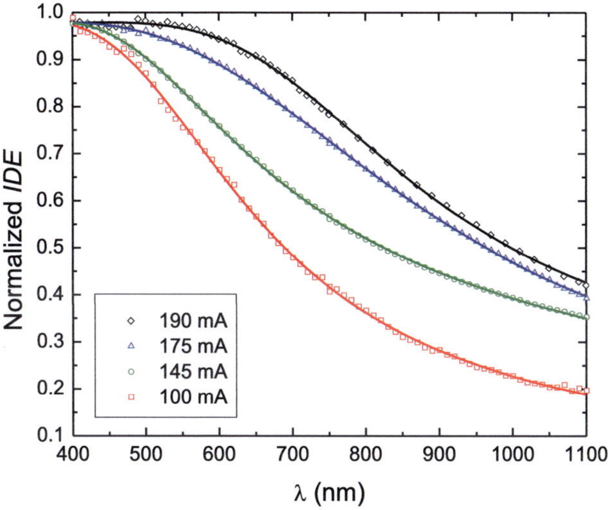

Figure 6.17: Normalized *IDE* spectra from the measurements at $I_B = 0.85 I_C^m$ for SNSPD made from NbN films with different I_{SP}. The solid lines are fits of eq. 6.5 to the data to extract the cut-off wavelength.

As an example, the normalized *IDE* for $0.85 I_C^m < I_B < 0.95 I_C^m$ are given in Fig. 6.17 together with the fit functions with which λ_0 was determined. It can now be seen that the cut-off wavelength is shifted continuously into the infrared range as the nitrogen content of the NbN film is reduced. Consequently, the highest λ_0 values are achieved for the film with $I_{SP} = 190 \, \text{mA}$.

From the minimum required photon energy given by eq. 2.6, the cut-off wavelength predicted by the hot-spot model can be formulated:

$$\lambda_C^{-1} = \frac{3\sqrt{\pi}}{4hc\varsigma} N_0 \Delta^2 w d \sqrt{D\tau_{th}} \left(1 - \frac{I_B}{I_C^d} \right). \tag{6.6}$$

The electronic density of states N_0 can be expressed in the framework of the free electron model as $N_0 = (R_S d D e^2)^{-1}$. The only parameter defined by the measurement conditions is the bias current I_B. We thus split equation 6.6 into the current-dependent part in the brackets and a prefactor κ_C, which is determined solely by material parameters and the wire geometry:

$$\lambda_C = \kappa_C \left(1 - \frac{I_B}{I_C^d} \right)^{-1} \tag{6.7}$$

$$\kappa_C = \frac{4hc\varsigma}{3\sqrt{\pi}} \frac{e^2 R_S}{\Delta^2 w} \sqrt{\frac{D}{\tau_{th}}}. \tag{6.8}$$

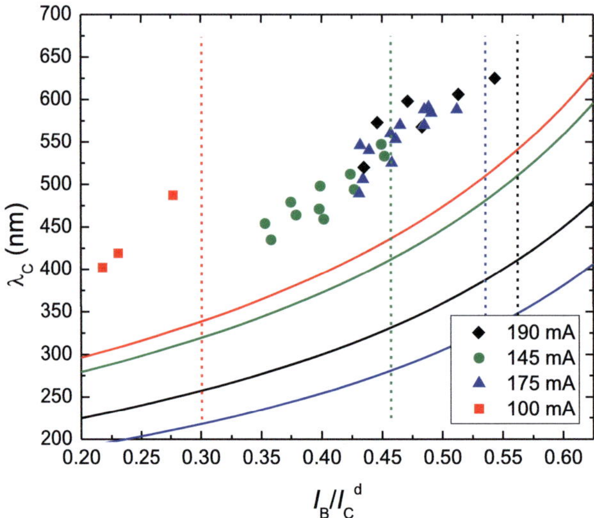

Figure 6.18: Cut-off wavelengths for all measured SNSPD over the bias current relative to the de-pairing value I_B/I_C^d at which the measurement was obtained. The color marks the deposition current of the NbN film from which the respective sample was structured. The dashed lines show the I_C^m/I_C^d values for each film, which marks the maximum at which the detector can be biased. The solid lines show the dependences predicted by the hot-spot model, with κ_C calculated from the material parameters.

To achieve large values for the cut-off wavelength, there are two requirements: A large pre-factor κ_C and a high ratio of I_C^m/I_C^d, which marks the limit of I_B/I_C^d in the bracket term. According to the measurements presented in chapter 3, the square resistance R_S decreases for increasing sputter currents, which leads to an lower value of κ_C. At the same time, the energy gap Δ decreases and the diffusion coefficient D increases, which both lead to a contrary effect. An unknown parameter is the quasiparticle multiplication efficiency ς. It is dependent on the photon energy [47] and, involving energy relaxation processes, most probably also material dependent. For further calculations we neglect those dependences and for all samples adopt the value $\varsigma = 0.43$, which was extracted from similar spectral measurement on SNSPD made from standard NbN [28]. The calculated values of κ_C, using the material parameters determined in the previous chapters, are given in table 6.2. The results show a minimum value for $I_{SP} = 175\,\text{mA}$ and the maximum value for $I_{SP} = 100\,\text{mA}$.

Figure 6.18 shows a plot of all measured cut-off wavelengths λ_C in dependence of the I_B/I_C^d ratio. Data points from samples that were structured from the same NbN film are shown in the same color. For each film the respective theoretical dependence with the calculated κ_C is shown by a solid line of the same color. Their dependence is qualitatively similar to the measured data, but with a shift to higher λ_C values. This off-set can be explained by the difference in definition of λ_C in the hot-spot model and the experimentally obtained λ_0. Remarkably, the experimental values do not show a significantly different behaviour for varying I_{SP}, but seem to follow the same dependence. In contrast to the expectation from the calculated κ_C values, the samples from the film with $I_{SP} = 100\,\mathrm{mA}$ achieve the lowest maximum cut-off wavelength. The reason for this is the small ratio of I_C^m/I_C^d in this film, which does not allow the samples to be biased at higher currents. The biasing limit I_C^m/I_C^d is marked in Fig. 6.18 by dashed vertical lines for all films in the respective colors. The samples from the film with $I_{SP} = 190\,\mathrm{mA}$ display the highest I_C^m/I_C^d value, and although the cut-off wavelengths measured at the same I_B/I_C^d are almost identical as for other films, it reaches the overall highest λ_0 values.

From this observations we conculde, that the impact of a material change on the cut-off wavelength is mainly given by the current-dependent term. The pre-factor κ_C depends on material properties that are interdependent and mainly favours the combination of high T_C and high R_S, both of which being properties that were used in the past to select materials for SNSPD application. However, for a material with greatly enhanced spectral bandwidth, a high κ_C value has to be paired with a high ratio I_C^m/I_C^d, because the current-dependent term diverges as $I_B/I_C^d \to 1$. For the material variation of the NbN films discussed here, this dependence dominates and the maximum cut-off wavelength increases as the nitrogen content in the NbN film is reduced.

Overall, SNSPD devices made from NbN films with reduced nitrogen content display enhanced detector performance as compared to the standard composition despite the lower T_C and R_S. At the same operation temperature $T = 4.2\,\mathrm{K}$ and bias current I_B, they show much lower DCR. The cut-off wavelength λ_C is increased due to an increased I_C^m/I_C^d ratio in the samples, leading to an improved DE in the infrared range. Additionally, an improved absorption coefficient in the films with increased niobium content (see section 3.5) benefits the application in SNSPD.

6.2.4 Temperature-dependent spectral detection efficiency

The temperature dependence of an SNSPD's spectral detection efficiency poses a remaining open question for the hot-spot model. A general observation of the various materials that have been used as SNSPD in the past revealed that materials with a smaller energy gap showed in general a broader spectral DE (compare section 2.1.2). Consequently, one should expect an improvement when the operation temperature is increased, since the temperature-dependent energy gap is then reduced. However, the experimental evidence is exactly opposite.

The first reported measurements at temperatures other than 4.2 K were aiming on reducing the dark count rate. Since the dark counts are thermally activated, lowering the operation temperature significantly improves on the DCR [159, 160]. Photon count rate measurements at varying temperatures on SNSPD from NbN [6, 156], Nb [42] and NbTiN [35] all report a significant improvement of DE as the temperature is decreased.

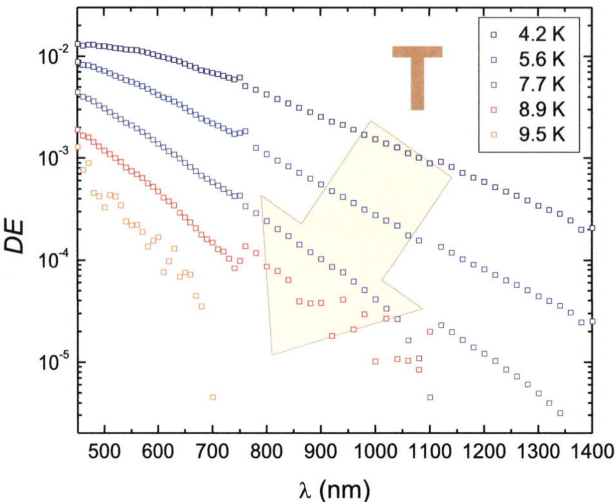

Figure 6.19: Temperature-dependent spectral detection efficiency of an SNSPD made from 4 nm thick NbN with standard composition and 100 nm linewidth. The measurements were taken at $0.95I_C^m(T)$ for various temperatures above 4.2 K.

Fig. 6.19 shows spectra of DE measured on one of the samples described in the last section, made from an NbN films with standard composition. The temperature was gradually increased up to 9.5 K, while the bias current was adjusted to $0.95I_C^m(T)$ for each temperature. As the temperature increases, the hot-spot plateau becomes shorter and the cut-off wavelength is reduced. For $T > 5.6$ K, λ_C is already in the UV and thus outside the measurement range. The measurements become increasingly noisy at higher temperatures, because of the rapid increase of the DCR. Above 9.5 K, the DCR becomes too strong for feasible measurements. Very similar measurements on SNSPD made from TaN thin films showed correspondent dependences when the operation temperature is reduced below 4.2 K [161].

If measurements are executed at constant ratios $I_B(T)/I_C^d(T)$ as those shown above, the temperature dependence of the critical current is irrelevant for the discussion. This leaves only the material parameters in κ_C according to eq. 6.8 to be considered. The dependence of the energy gap $\Delta(T)$ is given by eq. 3.7. It is almost constant for low temperatures and reduces rapidly to zero as T_C is approached. It's dependence alone can not describe the experimentally observed $\lambda_C(T)$ behaviour.

Another parameter, of which the temperature dependence was so far neglected, is the diffusion coefficient D. It describes the movement of the excited quasi-particles that were created after the photon absorption and have relaxed to energies slightly above the energy gap. Using the general relation $D = K_e/C_e$, where K_e is the electron thermal conductivity and C_e the electron specific heat capacity, the temperature dependence of D was calculated numerically in [161]. The resulting relation of the cut-off wavelength $\lambda_C \propto \Delta(T)^{-2}D(T)^{-0.5}$ shows an increase as $T \to 0$ which fits to the measurements. However, as the temperature is increased, after reaching a minimum the relation rises again. This is in the temperature region approaching T_C, where the strong dependence of $\Delta(T)$ becomes dominant. So far, this increase in λ_C as $T \to T_C$ could not be experimentally observed. It should also be noted that for $T \to T_C$, the assumption $n_S \approx \frac{N_0\Delta}{2}$ that was made in the deduction of the cut-off wavelength no longer holds. An accurate describtion of $\lambda_C(T)$ within the hot-spot model in this temperature range would therefore have to be reformulated.

The last two parameters, that could have a temperature dependence, influencing on λ_C are the quasiparticle multiplication efficiency ς and the thermalization time constant τ_{th}. However, these values can not be directly measured. Furthermore, they additionally depend on the energy of the absorbed photon. Correct modelling and inclusion of those parameter's dependence on temperature and photon wavelength could remove the remaining discrepancies between $\lambda_C(T)$ of the model and of the measurements.

6.3 SNSPD layout with optimized current distribution

The experimental investigation of the current crowding effect described in chapter 5 revealed deficits in the classical meander design. In particular at the end of each line, where a sharp $180°$ turn makes the connection to the next line, the current is condensed at the inner edge of the turn and locally reduces the critical current. While this effect reduces the overall critical current of the device, it also limits the maximum ratio of I_B/I_C^d that can be achieved in the straight portions of the detector. Since the detection efficiency depends on this factor, this means that only in the vicinity of the turns a high DE can be achieved, while the rest of the detector structure suffers from a reduced DE. In this section, an alternative detector layout is developed which aims to address this point with a homogenized current distribution. The results presented here were published in [162].

Since a certain detector area needs to be filled as densely as possible with the nanowire, it is not possible to omit turns. The turns should have no abrupt angles and as large radius of curvature as possible. The simplest structure fulfilling these requirements is a spiral. It consist only of one long turn where the radius reduces as it spirals inwards. To make the contacts for the bias current without an additional layer of structuring, a second spiral arm leads the nanowire out. Figure 6.20 shows such a spiral layout, where the contacts for the bias currents are marked by

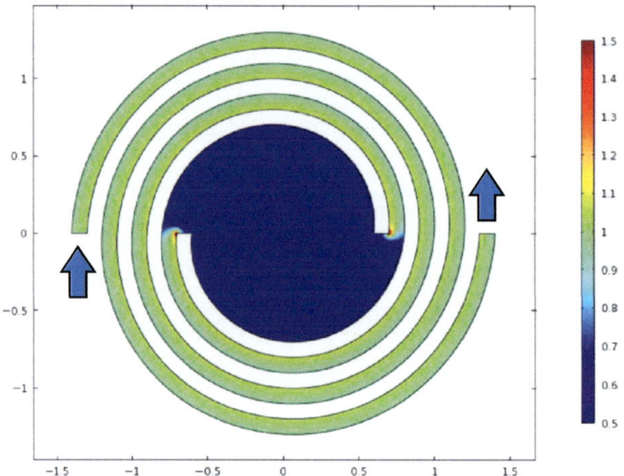

Figure 6.20: Simulation of the current distribution in a spiral detector layout. The color scale shows the current density with respect to the density of a straight wire of the same width (green), where red colors indicate increased and blue colors decreased density. The connections to the bias current are marked by the blue arrows.

blue arrows. In this case an Archimedean spiral was chosen, which is described by the following equations:

$$x = a\theta \cos\theta$$
$$y = a\theta \sin\theta, \tag{6.9}$$

where $\theta_{min} < \theta < \theta_{max}$ define the start and end point of the spiral. It has the special property that the distance (or pitch p) between two spiral arms is constant $p = a2\pi$. Because the spiral leading out will be in between the spiral leading in, the pitch (where pitch is the sum of wire width w and gap width g) of the final detector structure will be exactly half the pitch value of one spiral arm.

The color in Fig. 6.20 shows the current distribution simulation of such a structure according to the description of section 5.2.2. The spiral arms show a very homogeneous and continuous distribution with a slight increase (yellow) at the inner edge and a slight decrease (blue) at the outer edge. The weak spot of this design is obviously the central part: The decreasing radius of the spiral will lead to an increase of the current crowding effect in the center. This defines a lower limit for the radius of the spiral, depending on the maximum acceptable critical current reduction. The spiral can only start at this minimum radius, leaving an area where some structure must be placed that connects the two spiral arms and makes the change of the right turn to the left turn. The simulation shows that simply stopping the spiral arms leaves an effective 90° turn

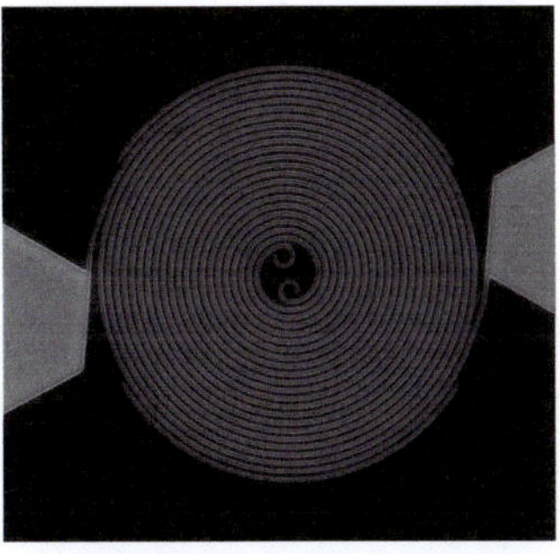 

Figure 6.21: Left: SEM image of 90 nm thick PMMA resist (black) on top of a 4 nm NbN film (light) after exposure with the electron beam and development. The image shows a spiral detector structure with 100 nm wire width and 60 nm gap width. Right: AFM image of one section of the same structure, after the layout of the resist was transfered into the NbN by reactive ion etching and the resist was stripped. Blue areas are the surface of the sapphire substrate (z-axis is exaggerated).

where the current density is strongly increased. One option to reduce this effect is to taper the spiral ends, so that the increased linewidth leads to a higher critical current in this region which counters the reduction by the current crowding.

For the central part of the spiral detector, the requirements of a high filling factor and large radius are contrary and the structure will necessarily have a lower filling factor than the spiral arms. Consequently, the detector will have a 'blind spot' in its center where the detection efficiency is lower than in the remaining parts. This can be especially disadvantageous for optical coupling to a Gaussian beam profile, where the maximum of the distribution is in the center.

6.3.1 Patterning of spiral nanowire detectors

The spiral detector structures were patterned by the same process as described in section 6.1.1. However, a few additional challenges required adjustment of the layout and of the process parameters in the electron-beam lithography part. The left picture in Fig. 6.21 shows an SEM image of the resist on top of a 4 nm thick NbN film after exposure with the electron beam and development. The black parts are remaining resist, while the light parts are the NbN surface which will be etched away in the next step.

With respect to the schematic shown above, the central structure was improved here by a continuous taper of the spiral end which increases the effective linewidth in this region to assure that it will not limit the critical current. Similarly, the outer ends of the spiral arms are tapered before they connect to the large contact areas at top and bottom. In this contact regions, one additional spiral branch on top and bottom reduces the proximity effect of the electron beam exposure and increases the homogeneity of the linewidth.

To achieve high uniformity in the linewidth of nanowire and gap over the whole length of the spiral proved to be challenging for the technological processes, especially when those linewidths are reduced below 100 nm. Due to the intrinsic cartesian coordinate system of the electron beam controller, round structures often suffer from step-like features that can lead either to constrictions in the wire or to short-cuts in the gap. The spiral design is more vulnerable to short-cuts than a meander, especially if they appear in the outer regions where they can cut off almost the whole detector. Instead of dividing the spiral structure into polygons that are exposed by multiple line scans, only a single line was exposed by the electron beam following the spiral curvature. The exposure dose was adjusted to yield the desired gap width of $g = 60$ nm. This procedure has lead to a significant improvement of the spread in the resistance values of the detector structures, indicating reduced number of short-cuts. High resolution atomic force microscopy on spiral detector structures has confirmed a high uniformity of the linewidth and a reduced number of remains in the gaps (see Fig. 6.21 (right)).

6.3.2 Current crowding effect in spiral detector structures

To examine whether the critical current reduction in spiral structures is indeed weaker than in a typical meander, test structures of both types are patterned on the same film and their critical currents are measured. The structures consist of a set of a spiral and a meander with identical wire width w and gap width g, as well as a straight bridge of the same width w as a reference. Three sets were made with w/g values of 300 nm/100 nm, 200 nm/100 nm and 100 nm/60 nm. Figure 6.22 shows SEM images of a few selected samples to illustrate the layout of the structures.

The critical current densities of several structures were measured at 4.2 K. The average values are shown in Fig. 6.23: For the 300 nm and 200 nm series, the meanders show a significantly reduced j_C with respect to the bridges, as expected. The spiral design, although also below the values of the straight bridge, reaches about 20% higher j_C values than the meander structure. The overall decrease of the critical current density with smaller linewidths is due to the proximity effect as described in chapter 3. A similar relative behaviour is observed in the 100 nm series, but here the reduction of j_C is weaker, although the smaller g value should lead to an even stronger reduction for the meander structures. Here, the spiral structure has only a 7% increase in j_C as compared to the meander, which already approaches the level of the statistical spread.

The measurements of the critical current confirm that the spiral detector layout suffers less from the current crowding effect than the meander. The difference seems to become smaller as w and g are reduced to values of typical SNSPD. However, this measurements only reveal the absolute reduction of j_C and do not allow any deduction on the current distribution in the structure. Even in two detectors with the same absolute j_C, the one with a more uniform distribution should yield a higher detection efficiency. To evaluate this, direct measurement of the detection efficiency are needed.

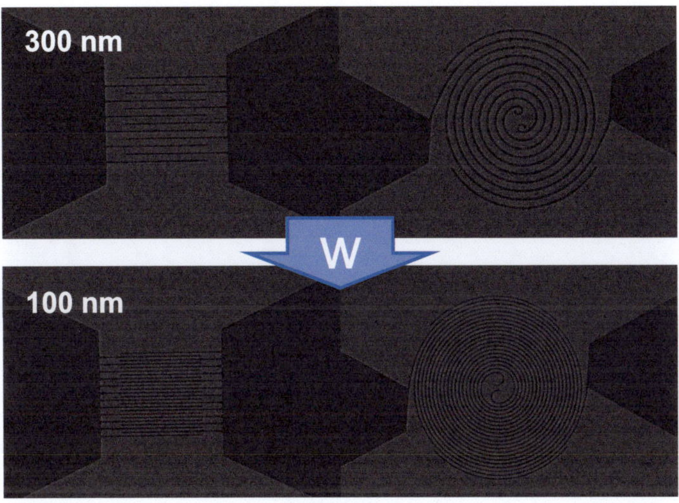

Figure 6.22: SEM images of four detector structures, where the left structures are meanders and the right structures spirals. The width was reduced from $w = 300\,\text{nm}$ (upper images) to $w = 100\,\text{nm}$ (lower images) while keeping the area approximately constant.

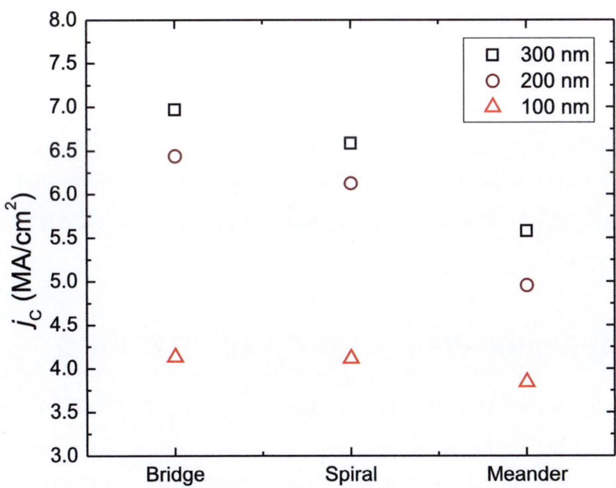

Figure 6.23: Critical current density j_C measured at $4.2\,\text{K}$ for spirals, meanders and straight lines with different nanowire width. All samples were structured on the same NbN film.

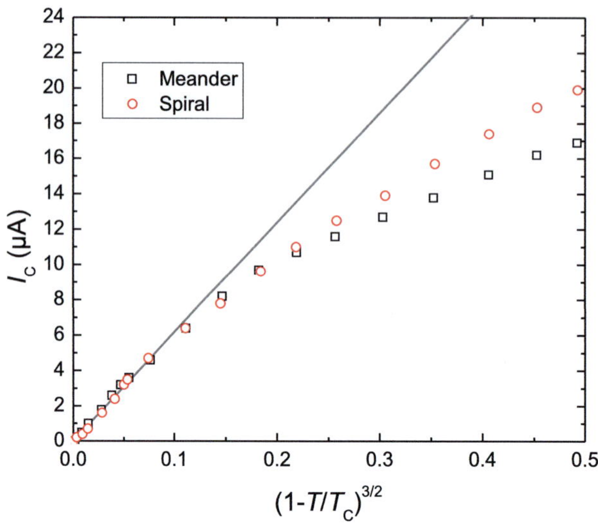

Figure 6.24: Temperature dependence of critical current I_C of a spiral and a meander structure with the same linewidth $w = 100\,$nm in the temperature scale $(1 - T/T_C)^{3/2}$ of the de-pairing critical current (solid line).

For nanowire widths of 200 nm and more the detection efficiency of SNSPD structures becomes too low for feasible measurements. The *DE* was thus determined only for two selected samples from the series with $w = 100\,$nm. Figure 6.24 shows the temperature-dependence of j_C of the samples, where the temperature is scaled in the dependence of the GL de-pairing critical current j_C^d. Both samples follow j_C^d for temperatures close to T_C and then deviate from it (grey line) and from each other. The j_C values of the meander stay below the values of the spiral in this region. This is the same behavior as described in chapter 5 and shows that the difference in $j_C(4.2\,$K$)$ of the samples originates from the reduced vortex penetration potential and not only from fabrication spread.

6.3.3 Spectral detection efficiency of a spiral SNSPD

The spectral detection efficiency was measured on one spiral and one meander with 100 nm width and 60 nm gap which were structured on the same 4 nm thick NbN film. The meander covered a detector area of $4.2 \times 4.2\,$µm. The spiral had an outer radius $r_a = 4.2$ µm and an inner inner radius $r_b = 0.6$ µm. The definition of the detector area is schematically illustrated in Fig. 6.25. The photon flux distribution is integrated over the area $A_{detector}$ as described in section 6.1.3. The inner blind spot is excluded from the definition, since it is not relevant for the purpose of this investigation. For a real application of this detector structure, the central part of the detector

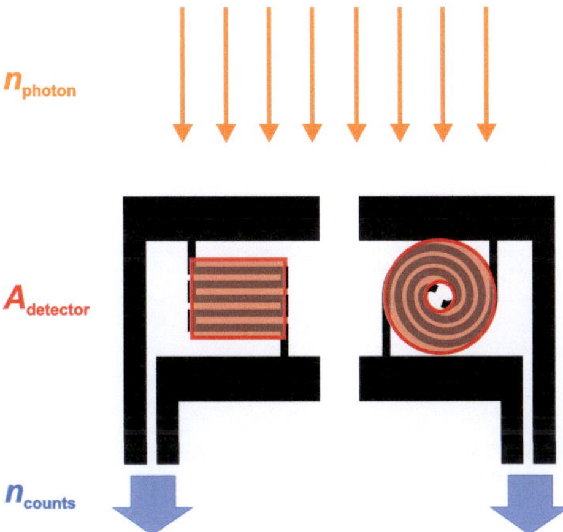

Figure 6.25: Schematic illustration for the definition of the detection efficiency for a spiral and meander layout. The number of counts n_{counts} registered by the readout are divided by the number of photons n_{photon} incident on the detector area $A_{detector}$.

would definitely have to be improved. In any case, inclusion of the central part into the definition here only changes the overall detection efficiency by less than 1 percentage point.

The spectral dependence of the measured detection efficiency for a spiral in direct comparison with a meander is shown in Fig. 6.26. Both measurements are performed at the same conditions, i.e. bias current $I_B = 0.95I_C$ and $T = 4.2$ K. The DCR was almost equal for both detector types. In the whole spectral range, the detection efficiency of the spiral is higher than for the meander. This is the case for all bias currents considered from 0.8 to $0.95I_C$ (not shown). Both structures show a similar dependence with a cut-off wavelength around 600 nm and a hot-spot plateau for smaller wavelengths. On this plateau, the meander reaches a DE of 18.2% and the spiral a clearly improved DE of 27.6%. In the infrared range, the spiral's DE is about 2.7 times the value of the meander.

To evaluate the cut-off wavelength of the intrinsic detection efficiency, the normalization method from section 6.1.4 is used. Each detector was measured at four bias currents from 0.8 to $0.95I_C$. The DE spectra are then normalized to the spectrum at $0.95I_C$. The three normalized detection efficiency spectra of the spiral SNSPD are shown in Fig. 6.27. The solid lines are fits of eq. 6.5 to extract the cut-off wavelengths λ_0 as fit parameters. The respective values are given in table 6.3. For all bias currents, the spiral detector has a higher λ_0, but only by about 15-20 nm. Figure 6.27 also reveals that the DE in the hot-spot regime, although approaching constant values, stay slightly below 1 and depend on the bias currents. This means, that the dependence $DE(I_B/I_C^m)$ is not yet in saturation for the considered currents. This behavior is common for NbN SNSPD at 4.2 K, contrary to e.g. TaN detectors, where for comparable conditions the saturation is reached.

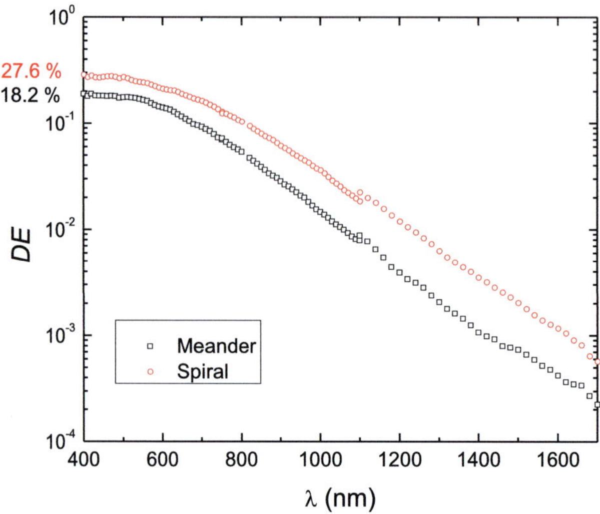

Figure 6.26: Spectral detection efficiency of spiral and meander detector structured from the same 4 nm thick NbN film with the same $w = 100$ nm and $g = 60$ nm values. The measurement was performed in both cases at $0.95I_C$ bias current and at $T = 4.2$ K.

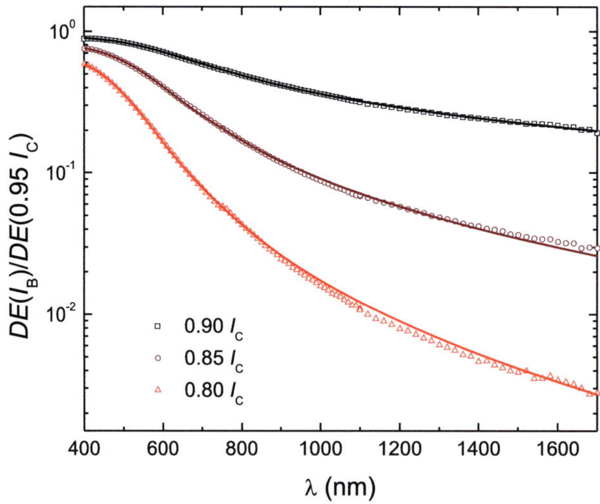

Figure 6.27: Spectral DE of a spiral detector measured at different I_B, normalized to the values at $0.95I_C$. The solid lines are fits of equation 6.5 to the data.

Table 6.3: Cut-off wavelength values λ_0 in nm for meander and spiral detector, evaluated for different bias currents given in brackets relative to I_C.

	Meander	Spiral
$\lambda_0(0.95I_C^m)$	620.0	632.8
$\lambda_0(0.90I_C^m)$	554.3	576.6
$\lambda_0(0.85I_C^m)$	494.0	525.3
$\lambda_0(0.80I_C^m)$	455.8	477.2

From the hot-spot model, the cut-off wavelength is expected to depend on the ratio of the bias current to the de-pairing critical current as described by eq. 6.7. Since both detectors are made from the same film, I_C^d is the same in both cases. However, the bias current applied is relative to the experimental critical current I_C^m, which is reduced below the de-pairing value by the mechanisms discussed in chapter 5. For the meander, I_C^m is lower than in the spiral, thus for a fixed bias factor I_B/I_C^m, e.g. 0.95, the fraction I_B/I_C^d is effectively lower and λ_C should be smaller than for the spiral operated at the same relative bias. Since the difference in I_C^m is small, the cut-off should be slightly reduced, which is exactly what is observed in the measurements.

The absolute detection efficiency on the other hand is not only defined by the overall I_C^m value, which can be limited by a single weak point in the structure, but by the distribution of the I_B/I_C^m ratio. In the meander, this ratio is high only in the turns at the side, because here I_C^m is low. The main part of the detecting area, however, consists of the straight line parts, where I_C^m is higher and thus I_B/I_C^m is lower. Consequently, these areas contribute a smaller local DE. If the ratio I_B/I_C^m is uniform over the whole detector area, a high DE will be provided by larger areas of the detector and the overall DE will be improved. In accordance to this, the spiral SNSPD shows a higher DE in the whole spectrum. The difference becomes apparent especially in the long-wavelength range, where the dependence $DE(I_B/I_C^m)$ is stronger. Although the absolute value of I_C^m is only slightly higher, the spiral layout seems to have a much more uniform current distribution due to reduced current crowding influence. The results are supported by similar observations made on NbTiN nanowires with a single turn [163].

A spiral detector layout has the added benefit of being less sensitive to the polarization of the incident light. For the classical meander design, the absorption probability depends on the polarization relative to the parallel nanowires [19, 154]. The wires act as a wire-grid polarizer: When the light is polarized parallel to the wires, the electric field is undisturbed, while for the perpendicular case the surface charges screen the field at the wire edges. The polarization-dependence of a spiral detector has been experimentally verified [35]. The detection efficiency to polarized 650 nm irradiation was measured on a NbTiN SNSPD with spiral layout. The degree of polarization (defined as n(parallel)-n(perpendicular)/n(total)) reduces from 0.28 for a meander structure down to only 0.03 for a spiral.

In both cases, the detection efficiency with respect to randomly polarized light is the mean value of the parallel and perpendicular detection efficiencies and no difference should be expected. Indeed the measurements reveal that in this case, the detection efficiency of the spiral is with 0.6% even lower than 1.8% for the meander. The reason for this is that in the design of the spiral, the turn in the center, where the inwards spiral changes to the second spiral going out, is once again structured as a sharp $180°$ turn. The critical current of the structure is limited by this turn, leaving the whole rest of the spiral with a low I_B/I_C^d ratio and consequently a low local detection efficiency. This observation further supports the assumption, that in a meander design the overall DE is reduced by the turns. For our spiral detectors, we can conclude from this investigation that the increase of detection efficiency is not caused by the change in polarization dependence, but seems to be indeed a consequence of the homogenized current distribution.

6.4 Summary

This chapter presented direct measurements on the performance of SNSPD devices. The samples are patterned from 4 nm NbN thin films in a combined electron-beam and photolithography process. The reactive ion-etching with SF_6 and Ar was optimized for the edge quality of the NbN nanowires. A cryogenic experimental setup was developed for measurements of the dark count rate and the spectral detection efficiency of the SNSPDs in the range $400 - 1700$ nm with an accuracy <4%. As a measure of the spectral bandwidth, the cut-off wavelength λ_C was extracted from the measurement data with a normalization method.

A detailed investigation of a series of SNSPDs structured from NbN films deposited at different sputter currents in vicinity of the transition from nitrogen to niobium phase was performed. The increase of the residual resistance ratio, the electron diffusion coefficient, and the decrease of the resistivity values of films deposited at high sputter current I_{Sp} indicate the growth of more uniform NbN films with a reduced amount of defects, which influences on the coherence length of the material. This results in a material dependence of the ratio between the measured critical current and the theoretically estimated depairing current, I_C^m/I_C^d, due to the reduction of the vortex penetration barrier by the current crowding effect. For the films with reduced nitrogen content, this value is increased up to 0.6. The I_C^m/I_C^d ratio can be identified as the strongest factor influencing on the spectral detection efficiency, as it defines the maximum value of the current-dependent factor in eq. 6.6 that can be achieved in the detector operation. For the maximum achievable cut-off wavelength of the detector, this limitation is much stronger than the material influence on the λ_C values predicted by the model. Thus, it can be asserted that the films with higher barrier for vortex penetration (larger I_C^m/I_C^d) are preferable for SNSPD development.

Consequently, the cut-off wavelength λ_C of SNSPD devices made from NbN films with reduced nitrogen content is shifted towards longer wavelengths by more than 25%, leading to a strongly improved infrared detection efficiency. Also, they demonstrate almost three orders of magnitude reduced DCR as compared to the reference sample with standard NbN composition. The steeper dependence of the DCR on the bias current further supports the assumption that the vortex penetration barrier is increased. At the same time, the detection efficiency on the hot-spot plateau reaches the same or even higher values than the SNSPDs with standard composition.

The cause for the strong temperature-dependence of λ_C remains an open point for further investigation. It can be partly explained by the consideration of the temperature dependence of the superconducting energy gap and the electronic diffusion coefficient. However, in this case an increase of λ_C in the vicinity of T_C is predicted which has so far not been observed experimentally. A more complete description of the problem should include the temperature dependence of the quasi-particle multiplication efficiency ς and the thermalization time constant τ_{th}, as well as their dependence on the photon energy.

To improve the uniformity of the current distribution in an SNSPD, a spiral detector layout was developed and devices were fabricated from NbN thin films. Measurements of the critical current showed that the new design suffers less from the reduction of I_C by the current crowding effect. For a line width of 300 nm, an increase of 20% in the critical current with respect to a conventional meander structure was achieved. However, as the width of the nanowire is reduced, the effect becomes weaker as predicted by theoretical models.

For structures with $w = 100$ nm and $g = 60$ nm, the spectral detection efficiency of the spiral shows only a slightly improved λ_C in accordance with the increase of the absolute critical current. However, the more uniform current distribution leads to a significant increase of the detection efficiency at operation conditions with identical DCR, especially at large wavelengths. For optical photons, the detection efficiency was improved by a factor of 1.5 from 18.2% to 27.6%. In the infrared range, at 1540 nm an even larger increase in the detection efficiency by a factor of 2.7 was achieved. The reduced dependence on polarization could be especially beneficial in quantum information experiments, where the polarization is often used as an additional degree of freedom. Additionally, the convenient optical coupling to incoming light beams due to the round shape should make the spiral SNSPD layouts attractive for many applications.

Based on the experimental results, further development of infrared SNSPD should focus on materials which are uniform, defect free and with high barrier for vortex penetration. The process for the patterning into sub-micrometer wide structures should provide uniform edges of high quality to exclude local reductions of the penetration potential. In addition, the layout of the SNSPD should include no sharp turns or steps to provide a homogeneous distribution of the critical current density.

7 Conclusion

In this thesis, several approaches for the improvement of Superconducting Nanowire Single-Photon Detectors (SNSPD) were studied. During the past decade, those devices have become competitive alternatives to photomultiplier tubes and avalanche photodiodes in the field of single-photon detection. Main considerations are given here to the advancement of the maximum count rates and the detection efficiency in the far-infrared spectral range, where SNSPDs exhibit considerable advantages over semiconductor devices.

The spectral bandwidth of the detection efficiency is limited by an exponential decay once the cut-off wavelength is surpassed. This threshold can be improved by optimization of the material properties of the superconductor. Such material properties were studied in the first part of this work on the basis of NbN thin films. The reactive magnetron sputtering process for the deposition of films with thickness in the range of $3-30\,\mathrm{nm}$ allows the continuous variation of the NbN stoichiometry. A systematic change in material properties around the standard composition with the maximum T_C value was realized by careful control of sputter conditions. A general transition from metal-like to isolating was observed in the resistive properties as the nitrogen content was increased, which is mainly attributed to the decreasing conductivity of the grain boundaries. To achieve a low number of electronic states, it would thus be favourable to increase the nitrogen content of the NbN films. However, this dependence is contrary to other factors: The electron diffusion coefficient and the coherence length of the NbN thin films extracted by measurements of the second critical magnetic field, both increase as the nitrogen content is reduced. The temperature-dependent energy gap was measured directly on $4\,\mathrm{nm}$ thick NbN films by THz spectroscopy. A reduced energy gap would lead to a larger number of quasiparticles genereated by the photon's energy and is found for the films with reduced nitrogen content. Lastly, the films were investigated by ellipsometry and their optical properties were extracted on the basis of a multi-layer model. The absorption probability of the films profits from the increased number of charge carriers for the films with reduced nitrogen content. For longer wavelengths, the absorptance generally increases further and the dependence on the material composition becomes even stronger.

The maximum count rate of an SNSPD is defined by the relaxation of the voltage pulse after a detection event has been triggered. While this is in the standard meander-shaped design mostly dominated by the device inductance, improved designs for high count rates are readily available. The physical limitation is given by the intrinsic relaxation processes taking place in the superconducting film upon absorption of a photon. Those processes were investigated by a frequency-domain measurement system, which allows the continuous variation of the excitation frequency in a broad range. From the modulation frequency dependence of the response signal, the energy relaxation time constant of NbN thin films was extracted. It is mainly influenced by the electron-phonon interaction time of the material and the escape time of phonons to the substrate, which dominates for large film thicknesses. Consequently, the energy relaxation time constant is minimized for thin films. Measurements of the material dependence reveal, that a further reduction of the relaxation time is possible for films with increased nitrogen content. However, for all considered chemical compositions, the relaxations are fast enough to allow high count rates. In the case of 4 nm thick films the values are around 100 ps, which would make count rates of >1 GHz feasible.

The next apporach to improve the spectral bandwidth of an SNSPD was to increase the ratio of the bias current to the depairing critical current. Measurements on single bridge structures of varying dimensions and materials show, that the de-pairing critical current is reached near T_C. At lower temperatures, the critical current is reduced by the penetration and movement of magnetic vortices. The range in which the de-pairing current can be reached is increased as the nanowire dimensions are reduced, because the self-field created by the current in the nanowire is then smaller. The same effect as an increase of the bridge width is observed by application of an external magnetic field. For niobium wires with $d < 13$ nm and $w < 400$ nm, j_C^d can be reached in almost the full temperature range from 4.2 K to T_C. As an alternative material with low self-field and high pinning forces, the ferroarsenide superconductor Ba-122 was investigated. A structuring process was developed, that allowed the patterning of sub-µm bridges. The results of the measurements were the first reports on the critical current of structured Ba-122 samples. In contrast to Nb, the material displays a wide range of temperatures in which j_C^d can be reached, even for $d = 50$ nm and $w > 1$ µm. For SNSPD with special requirements on structure sizes, Ba-122 could thus be a potential material of choice. For conventional SNSPD materials like NbN and TaN and typical nanowire dimensions $d = 4$ nm and $w < 100$ nm, the ratio j_C^m / j_C^d at 4.2 K can be very close to one for straight wires with ideal edges. The edge quality defined by the structuring process influences strongly on the height of the barrier for vortex penetration. A quantitative description of the penetration barrier and its influencing factors remains the main open challenge for further improvement of the straight nanowire critical current.

Another aspect that has very recently attracted attention is the reduction of the critical current by the geometry of the device. The accumulation of the current at the inner edge of a bend reduces the vortex penetration barrier and thus causes a switching to the normalcunducting state before the critical current would have been reached in the respective straight part of the wire. Especially in the widely used meander shaped design of an SNSPD, with the sharp turns at the end of each line, this leads to a reduction in the overall critical current and to a strongly inhomogeneous local ratio j_B / j_C^d. This current crowding effect was already known for some time, but was recently expanded to give quantitative predictions for the special case of nanowire dimensionalities. For a first experimental verification of this effect, the critical current of nanowires with bends was investigated on samples structured from NbN and Nb thin films. In both cases, a gradual reduction of j_C with increasing angle was observed. The reduction, however, was less pronounced than theoretically predicted. It was further verified, that the current crowding effect can be countered by an external magnetic field with correct polarity if the structure has turns in only one direction. In this case, the application of the field leads to an effective increase in the critical current. This effect could in principle be used to improve the SNSPD performance, but a special device layout would be neccessary.

Based on the above results, approaches to improve the SNSPD detector performance were formulated and respective devices were structured from the NbN thin films to directly measure the device performance. A fiber-based cryogenic setup was developed that allows the measurement of the detection efficiency for wavelengths $400 - 1700$ nm. From the spectral dependence, the cut-off wavelength of the intrinsic detection efficiency was determined.

The variation of the chemical composition of NbN revealed, that in contrast to the expectations from the material dependences, both the dark count rate as well as the spectral dependence can be clearly improved by reduction of the nitrogen content in the NbN films. The dominating factor for the improvement of the cut-off wavelength was identified to be the ratio of I_C^m/I_C^d at 4.2 K, which rises strongly as the nitrogen content is reduced. Compared to the standard composition, the cut-off wavelength could be increased by more than 20%, while at the same time the dark count rate is reduced by almost three orders of magnitude. The dependence of the cut-off wavelength on the bias current can be qualitatively well described within the hot-spot model. The temperature-dependence, however, behaves opposite to the expectations: An increase of the SNSPD operation temperature strongly reduces the spectral bandwith of the detection efficiency, although the energy gap is reduced. This conflict can be partly resolved by taking into account the temperature dependence of the quasiparticle diffusion coefficient. Further improvements of the model would have to include additional temperature-dependences, for example of the quasiparticle multiplication efficiency and the thermalization time constant.

To minimize the current crowding effect, a spiral SNSPD layout was developed. The absolute gain in the critical current density is significant for relatively large nanowire widths, but decreases for narrower wires. For $w = 100$ nm, the critical current of a spiral SNSPD structure is only increased by $\approx 7\%$ with respect to a meander structure. Consequently, the cut-off wavelength of the intrinsic detection efficiency of the spiral device is improved only slightly, in accordance with the hot-spot model. Nevertheless, the improved homogeneity of the local ratio j_C^m/j_C^d leads to a larger absolute detection efficiency of the device. In the optical range, the detection efficiency of the spiral structure reaches 27.6% (compared to 18.2% of the meander reference sample), which is already very close to the absorption limit of $\approx 30\%$. Combined with the larger cut-off wavelength, this leads to an effective increase of the detection efficiency in the infrared range by a factor of >2.5. An additional benefit is the low sensitivity of the spiral's detection efficiency to the polarization of the incident light, making these devices especially interesting for applications in optical communications and quantum experiments.

List of own publications

Reviewed journals

U. Pracht, E. Heintze, C. Clauss, D. Hafner, R. Bek, D. Werner, S. Gelhorn, M. Scheffler, M. Dressel, D. Sherman, B. Gorshunov, K. Il'in, **D. Henrich**, and M. Siegel. Electrodynamics of the Superconducting State in Ultra-Thin Films at THz Frequencies. *IEEE: Transactions on Terahertz Science and Technology*, 3:269, 2013.

D. Henrich, L. Rehm, S. Dörner, M. Hofherr, K. Ilin, A. Semenov, and M. Siegel. Detection Efficiency of a Spiral-Nanowire Superconducting Single-Photon Detector. *IEEE: Transactions on Applied Superconductivity*, 23:2200405, 2013.

A. Engel, K. Inderbitzin, A. Schilling, R. Lusche, A. Semenov, H.-W. Hübers, **D. Henrich**, M. Hofherr, K. Il'in, and M. Siegel. Temperature-dependence of detection efficiency in NbN and TaN SNSPD. *IEEE: Transactions on Applied Superconductivity*, 23:2300505, 2013.

M. Hofherr, M. Arndt, K. Il'in, **D. Henrich**, M. Siegel, J. Toussaint, T. May, and H.-G. Meyer. Time-Tagged Multiplexing of Serial Biased Superconducting Single-Photon Detectors. *IEEE: Transactions on Applied Superconductivity*, 23:2501205, 2013.

L. Rehm, **D. Henrich**, M. Hofherr, S. Wünsch, P. Thoma, A. Scheuring, K. Ilin, M. Siegel, S. Haindl, K. Iida, F. Kurth, B. Holzapfel, and L. Schultz. Infrared photo-response of Fe-shunted Ba-122 thin film microstructures. *IEEE: Transactions on Applied Superconductivity*, 23:7501105, 2013.

D. Henrich, S. Dörner, M. Hofherr, K. Il'in, A. Semenov, E. Heintze, M. Scheffler, M. Dressel, and M. Siegel. Broadening of hot-spot response spectrum of superconducting NbN nanowire single-photon detector with reduced nitrogen content. *Journal of Applied Physics*, 112:074511, 2012.

M. Hofherr, O. Wetzstein, S. Engert, T. Ortlepp, B. Berg, K. Ilin, **D. Henrich**, R. Stolz, H. Toepfer, H.-G. Meyer, and M. Siegel. Orthogonal sequencing multiplexer for superconducting nanowire single-photon detectors with RSFQ electronics readout circuit. *Optics Express*, 20:28683, 2012.

D. Henrich, P. Reichensperger, M. Hofherr, K. Il'in, M. Siegel, A. Semenov, A. Zotova, and D.Yu. Vodolazov. Geometry-induced reduction of the critical current in superconducting nanowires. *Physical Review B*, 86:144504, 2012.

T. May, J. Toussaint, R. Grüner, M. Schubert, H.-G. Meyer, B. Dietzek, J. Popp, M. Hofherr, K. Il'in, **D. Henrich**, M. Arndt, and M. Siegel. Superconducting nanowire single-photon detectors for picosecond time resolved spectroscopic applications. *Biophotonics: Photonic Solutions for Better Health Care III, Proceedings of the SPIE*, 8427:84274B, 2012.

K. Il'in, **D. Rall**, M. Siegel, A. Semenov. Critical current density in thin superconducting TaN film structures. *Physica C: Superconductivity*, 479:176 - 178, 2012.

D. Rall, L. Rehm, K. Il'in, M. Siegel, K. Iida, S. Haindl, F. Kurth, B. Holzapfel, L. Schultz, J. Yong, T. Lemberger. Penetration and de-pinning of vortices in sub-micrometer $Ba(Fe,Co)_2As_2$ thin film bridges. *Physica C: Superconductivity*, 479:164 - 166, 2012.

M. Hofherr, **D. Rall**, K. Il'in, A. Semenov, H.-W. Hübers, and M. Siegel. Dark Count Suppression in Superconducting Nanowire Single Photon Detectors. *Journal of Low Temperature Physics*, 167(5):822-826, 2012.

K. Ilin, M. Hofherr, **D. Rall**, M. Siegel, A. Semenov, A. Engel, K. Inderbitzin, A. Aeschbacher, and A. Schilling. Ultra-thin TaN Films for Superconducting Nanowire Single-Photon Detectors. *Journal of Low Temperature Physics*, 167:809-814, 2011.

T. Ortlepp, M. Hofherr, L. Fritzsch, S. Engert, K. Il'in, **D. Rall**, H. Töpfer, H.-G. Meyer, and M. Siegel. Demonstration of digital readout circuit for superconducting nanowire single photon detector. *Optics Express*, 19:18593-601, 2011.

P. Probst , A. Scheuring, M. Hofherr, **D. Rall**, S. Wünsch, K. Il'in, M. Siegel, A. Semenov, A. Pohl, H.-W. Hübers, V. Judin, A.-S. Müller, A. Hoehl, R. Müller, and G. Ulm. $YBa_2Cu_3O_{7-x}$ quasioptical detectors for fast time-domain analysis of THz synchrotron radiation. *Applied Physics Letters*, 98:043504, 2011.

D. Rall, K. Il'in, K. Iida, S. Haindl, F. Kurth, T. Thersleff, L. Schultz, B. Holzapfel, and M. Siegel. Critical current densities in ultra-thin $Ba(Fe,Co)_2As_2$ microbridges. *Physical Review B*, 83:134514, 2011.

M. Hofherr, **D. Rall**, K. S. Il'in, M. Siegel, A. Semenov, H.-W. Hübers, and N. Gippius. Intrinsic detection efficiency of superconducting nanowire single-photon detectors with different thicknesses. *Journal of Applied Physics*, 108:014507, 2010.

M. Hofherr, **D. Rall**, K. S. Il'in, A. Semenov, N. Gippius, H.-W. Hübers, and M. Siegel. Superconducting nanowire single-photon detectors: Quantum Efficiency vs. thickness of NbN films. *Journal of Physics: Conference Series*, 234(1):012017, 2010.

K. Il'in, **D. Rall**, M. Siegel, A. Engel, A. Schilling, A. Semenov, and H.-W. Hübers. Influence of thickness, width and temperture on critical current density of Nb thin film structures. *Physica C: Superconductivity*, 470:953-956, 2010.

D. Rall, P. Probst, M. Hofherr, S. Wünsch, K. Il'in, U. Lemmer, and M. Siegel. Relaxation time in NbN and YBCO thin films under optical irradiation. *Journal of Physics: Conference Series*, 234(4):042029, 2010.

Conferences

D. Henrich, P. Reichensperger, Y. Luck, L. Rehm, S. Dörner, J. M. Meckbach, M. Hofherr, K. Il'in, M. Siegel, A. Semenov, H.-W. Hübers, and D. Vodolazov. Effect of Bends on the Critical Current in Superconducting Nanowires. *Applied Superconductivity Conference*, Portland, USA (2012).

D. Henrich, P. Reichensperger, M. Hofherr, K. Il'in, M. Siegel, A. Semenov, H.-W. Hübers, and D. Vodolazov. Influence of geometry on the critical current of superconducting nanowire single-photon detectors. *5th Dresden-Karlsruhe Seminar on Materials and Applications in Superconductivity*, Bad Liebenzell, Germany (2012).

D. Henrich, P. Reichensperger, M. Hofherr, K. Il'in, M. Siegel, A. Semenov, H.-W. Hübers, and D. Vodolazov. Influence of geometry on the critical current of superconducting nanowire single-photon detectors. *DPG Frühjahrstagung*, Berlin, Germany, (2012).

D. Rall, L. Rehm, K. Il'in, M. Siegel, K. Iida, S. Haindl, F. Kurth, B. Holzapfel, and L. Schultz. Penetration and de-pinning of vortices in sub-micrometer $Ba(Fe,Co)_2As_2$ thin film bridges. *Seventh international conference on Vortex matter in nanostructured superconductors (VORTEX VII)*, Rhodes, Greece (2011).

D. Rall, J. Toussaint, M. Hofherr, S. Wuensch, K. S. Il'in, A. Semenov, H.-W. Huebers, and M. Siegel. Detection Efficiency and Cut-off Wavelength of NbN-SNSPD with Different Stoichiometry. *14th International Workshop on low temperature detectors*, Heidelberg, Germany (2011).

D. Rall, M. Hofherr, J. Toussaint, K. S. Il'in, and M. Siegel. Energy relaxation processes in SNSPDs. *4th Dresden-Karlsruhe Seminar on Materials and Applications in Superconductivity*, Bad Schandau, Germany (2011).

D. Rall. Superconducting Nanowire Single Photon Detectors: Improvement of spectral bandwidth by material variation. *Karlsruhe Days of Optics and Photonics*, Karlsruhe, Germany (2011).

D. Rall, K. Il'in, M. Siegel, K. Iida, S. Haindl, F. Kurth, T. Thersleff, L. Schultz, and B. Holzapfel. Transport properties of thin Ba(Fe,Co)$_2$As$_2$ film microbridges. *DPG Frühjahrstagung*, Dresden, Germany (2011).

D. Rall, P. Probst, M. Hofherr, S. Wünsch, K. Il'in, S. Haindl, K. Iida, U. Lemmer, and M. Siegel. Investiagtion of Energy Relaxation Processes in ultra-thin Superconducting Films with Frequency Domain Measurements. *3rd Dresden-Karlsruhe Seminar on Materials and Applications in Superconductivity*, Bad Liebenzell, Germany (2010).

D. Rall, J. Toussaint, M. Hofherr, S. Wünsch, K. Il'in, A. Semenov, H.-W. Hübers, U. Lemmer, and M. Siegel Improvement of SNSPD Detection Efficiency by Variation of NbN Chemical Composition. *Tagung kryoelektronische Bauelemente*, Zeuthen, Germany (2010).

D. Rall, K. S. Il'in, M. Hofherr, P. Probst, S. Wünsch, U. Lemmer, M. Siegel, A. D. Semenov, and H.-W. Hübers. Investigation of NbN thin film response to optical irradiation for detector application. *2nd S-Pulse Karlsruhe Detector Workshop*, Karlsruhe, Germany (2010).

D. Rall, K. Il'in, M. Siegel, and U. Lemmer. Investigation of Energy Relaxation Processes in Ultra-Thin NbN Films with Frequency Domain Measurements. *Tagung Kryoelektronische Bauelemente*, Oberhof, Germany (2009).

D. Rall. Superconducting Radiation Detectors: Investigation of Energy Relaxation Processes in Ultra-Thin NbN Films with Frequency Domain Measurements. *Karlsruhe Days of Optics and Photonics*, Karlsruhe, Germany (2009).

D. Rall, P. Probst, M. Hofherr, S. Wünsch, K. Ilin, U. Lemmer, and M. Siegel. Energy Relaxation Time in NbN and $YBa_2Cu_3O_{7-x}$ Thin Films under Optical Radiation. *European Conference on Applied Superconductivity*, Dresden, Germany (2009).

D. Rall. Energy Relaxation in Thin Superconducting Films. *2nd Dresden-Karlsruhe Seminar on Materials and Applications in Superconductivity*, Colditz, Germany (2009).

D. Rall, M. Hofherr, K. Il'in, S. Wünsch, M. Siegel, U. Lemmer, A. Semenov, and H.-W. Hübers. Investigation of Energy Relaxation Processes in NbN thin films using Optical Irradiation. *DPG Frühjahrstagung*, Dresden, Germany, (2009).

D. Rall. Properties of NbN thin films for application in superconducting single photon detectors. *1st Dresden-Karlsruhe Seminar on Materials and Applications in Superconductivity*, Bad Liebenzell, Germany (2008).

D. Rall, K. Il'in, S. Wunsch, F. Glockler, U. Lemmer, and M. Siegel. Response of NbN ultra-thin Superconducting Films to Optical Irradiation. *DPG Frühjahrstagung*, Berlin, Germany (2008).

List of Figures

List of Tables

List of Equations

Nomenclature

A	Absorptance
A_{det}	Detector area
a_0	Lattice constant
α	Absorption coefficient or angle of nanowire curvature or phonon interface transparency
ABS	Absorption probability
AFM	Atomic force microscopy
B_{C1}	Lower critical magnetic flux density
B_{C2}	Upper critical magnetic fulx density
B_{edge}	Magnetic field created at the edge of a nanowire by the transport current
Ba-122	$BaCo_2As_2$
BCS	Bardeen-Cooper-Schrieffer theory of superconductivity
c	Speed of light
C_e	Electron specific heat
C_p	Phonon specific heat
d	Film thickness
D	Electron diffusion coefficient
d_S	Thickness of remaining effective superconducting film
d_N	Thickness of normal conducting surface layer
δ	Skin depth
Δ	Superconducting energy gap or phase difference between polarization axes in ellipsometry context
ΔT	Temperature span of superconducting transition
DCR	Dark count rate
DE	Detection efficiency
e	Elementary charge
E	Electric field
E_p/E_s	Electric filed component polarized parallel/ perpendicular to plane of incidence
E_0	Minimum photon energy that leads to hot-spot formation
E_{ph}	Photon energy
EELS	Electron energy loss spectroscopy

$\tilde{\varepsilon}$	Complex dielectric function
ε_0	Energy scale parameter of vortex nucleation
$\varepsilon_1/\varepsilon_2$	Real/ imaginary component of dielectric function
f	Frequency of radiation
f_ε	cut-off frequency
F_L	Lorentz Force
FDTD	Finite-difference time-domain method
FWHM	Full-width half maximum
g	Width of gap between two meander lines
GL	Ginzburg-Landau theory
h	Planck's constant or distance between fiber end and sample surface
$\hbar$	Reduced Planck's constant
H_{C1}	Lower critical magnetic field
H_{C2}	Upper critical magnetic field
I_B	Applied bias current
I_C	Critical current
$I_{C,max}$	Maximum critical current in magnetic field dependence
I_C^d	De-pairing critical current
I_C^m	Measured critical current
I_{SP}	Discharge current of sputter deposition
I_s/I_p	Light intensity polarized perpendicular/ parallel to analyzer
I_0	Current scale parameter of vortex nucleation
IDE	Intrinsic detection efficiency
j_S	Average supercurrent density
j_C	Critical current density
j_C^d	Depairing critical current density
j_C^p	Depinning current density
j_C^m	Measured critical current density
j_H	Hysteresis current density
j_{pen}	Penetration current density
j_{tr}	Transport current density
k	Imaginary component of index of refraction
k_B	Boltzmann's constant
K_e	Electron thermal conductivity
$\kappa_{C,0}$	Material-dependent prefactor of cut-off wavelength
l	Electron mean free path
λ	wavelength of radiation or magnetic penetration depth
Λ	Pearl length
λ_C	Cut-off wavelength of hot-spot model

λ_{eff}	Effective thin film penetration depth
λ_L	London penetration depth
λ_0	Empirical cut-off wavelength
L-He	Liquid Helium
LSAT	$(\text{La,Sr})(\text{Al,Ta})\text{O}_3$
M	Multiplication factor of electron relaxation avalanche process
μ_0	vacuum permeability
n	Real component of index of refraction or total count rate
$\tilde{n}$	Index of refraction
n_e	Concentration of non-equilibrium quasiparticles
N_e	Number of non-equilibrium quasiparticles
$N_{e,c}$	Number of required non-equilibrium quasiparticles for hot-spot formation
n_{ph}	Rate of phonons incident on detector area
n_S	Mean Cooper pair density
N_0	Electron density of states
ν	frequency of radiation
v_S	Phonon speed in superconducting film
OCE	Optical coupling efficiency
ω	Angular frequency of radiation or modulation frequency
Ω	Attempt rate of vortex penetration
ω_D	Debye frequency
p	Pitch of spiral
P	Optical power or probability
p_{Ar}	Partial pressure of Argon gas
P_{det}	Optical power incident on detector area
P_{in}	Radiation power coupled into the system
P_{opt}	Incident optical power
P_{out}	Total optical power emitted by fiber end
p_{N2}	Partial pressure of Nitrogen gas
P_0	Unmodulated incident radiation power
PLD	Pulsed laser deposition
PMMA	Polymethyl methacrylate
PMT	Photomultiplier tube
Φ_0	Magnetic flux quantum
Ψ	Angle between long axis of ellipse and s-polarization
$\Psi(x)$	Digamma function
r	Distance from photon absorption site or radius of nanowire curvature
r_s/r_p	Complex amplitude of s-polarized/ p-polarized reflected light
r_{dep}	Deposition rate of sputter process

R	Reduction factor of current crowding effect
$R_{(x)}$	Resistance (at temperature x)
r_a/r_b	Outer/ inner radius of spiral design
R_N	Normal state resistance of superconducting material
R_S	Square resistance of film
RRR	Residual resistivity ratio
$\rho_{(x)}$	Resistivity (at temperature x)
ρ_{int}	Resistivity of interface between normal conductor and superconductor
RIE	Reactive ion etching
SDE	System detection efficiency
SEM	Scanning electron microscopy
SNAP	Superconducting nanowire avalanche photodetector
SNSPD	Superconducting Nanowire Single-Photon Detector
SPAD	Single-photon avalanche photodiode
Sr-122	$SrCo_2As_2$
ς	Efficiency of quasiparticle multiplication process
t	Time
T	Temperature
T_C	Critical temperature of superconductor
T_C^0	Ideal critical temperature without reduction by proximity effect
T_{dev}	Temperature of deviation of measured critical current from de-pairing critical current
T_e	Effective temperature of electron subsystem
T_p	Effective temperature of phonon subsystem
T_0	Bath temperature
θ	Angle of beam reflection or spiral parameter
$\theta_{min}/\theta_{max}$	Minimum/ maximum spiral parameter
τ	Scattering/ relaxation time constant
τ_{ee}	Electron-electron scattering time constant
τ_{ep}	Electron-phonon scattering time constant
τ_{es}	time constant of phonon escape to substrate
τ_{pe}	Phonon-electron scattering time constant
τ_{th}	time constant of quasiparticle thermalization
τ_ε	Energy relaxation time constant
TCSPC	Time-correlated single-photon counting
TDGL	Time-dependent Ginzburg-Landau equations
TEM	Transmission electron microscopy
U	Voltage or energy potential
U_{trig}	Trigger level of pulse counter

V	Volume
v_C	Critical Cooper pair velocity
v_S	Mean Cooper pair velocity
v_N	Mean electron velocity in normal state
VAP	Vortex-antivortex pair
w	width of nanowire
XRD	X-ray diffractometry
ξ	Effective coherence length
ξ_{GL}	Ginzburg-Landau coherence length
ξ_0	BCS coherence length

Bibliography

[1] C. M. Natarajan, M. G. Tanner, and R. H. Hadfield. "Superconducting nanowire single-photon detectors: physics and applications". *Superconductor Science and Technology*, **25**, 063001, 2012.

[2] S. Cova, M. Ghioni, A. Lotito, I. Rech, and F. Zappa. "Evolution and prospects for single-photon avalanche diodes and quenching circuits". *Journal of Modern Optics*, **51**, 1267–1288, 2004.

[3] M. A. Itzler, X. Jiang, M. Entwistle, K. Slomkowski, A. Tosi, F. Acerbi, F. Zappa, and S. Cova. "Advances in InGaAsP-based avalanche diode single photon detectors". *Journal of Modern Optics*, **58**, 174–200, 2011.

[4] R. H. Hadfield. "Single-photon detectors for optical quantum information applications". *Nature Photonics*, **3**, 696–705, 2009.

[5] R. Romestain, B. Delaet, P. Renaud-Goud, I. Wang, C. Jorel, J.-C. Villegier, and J.-P. Poizat. "Fabrication of a superconducting niobium nitride hot electron bolometer for single-photon counting". *New Journal of Physics*, **6**, 129, 2004.

[6] G. Gol'tsman, O. Minaeva, A. Korneev, M. Tarkhov, I. Rubtsova, A. Divochiy, I. Milost-naya, G. Chulkova, N. Kaurova, B. Voronov, D. Pan, J. Kitaygorsky, A. Cross, A. Pearlman, I. Komissarov, W. Slysz, M. Wegrzecki, P. Grabiec, and R. Sobolewski. "Middle-Infrared to Visible-Light Ultrafast Superconducting Single-Photon Detectors". *Applied Superconductivity, IEEE Transactions on*, **17**, 246 –251, june 2007.

[7] E. A. Dauler, A. J. Kerman, B. S. Robinson, J. K. Yang, B. Voronov, G. Goltsman, S. A. Hamilton, and K. K. Berggren. "Photon-number-resolution with sub-30-ps timing using multi-element superconducting nanowire single photon detectors". *Journal of Modern Optics*, **56**, 364–373, 2009.

[8] K. Inderbitzin, A. Engel, A. Schilling, K. Il'in, and M. Siegel. "An ultra-fast superconducting Nb nanowire single-photon detector for soft x-rays". *Applied Physics Letters*, **101**, 162601, 2012.

[9] M. Rosticher, F. R. Ladan, J. P. Maneval, S. N. Dorenbos, T. Zijlstra, T. M. Klapwijk, V. Zwiller, A. Lupascu, and G. Nogues. "A high efficiency superconducting nanowire single electron detector". *Applied Physics Letters*, **97**, 183106, 2010.

[10] K. Suzuki, S. Miki, Z. Wang, Y. Kobayashi, S. Shiki, and M. Ohkubo. "Superconducting NbN Thin-Film Nanowire Detectors for Time-of-Flight Mass Spectrometry". *Journal of Low Temperature Physics*, **151**, 766–770, 2008.

[11] K. Suzuki, S. Miki, S. Shiki, Y. Kobayashi, K. Chiba, Z. Wang, and M. Ohkubo. "Ultrafast ion detection by superconducting NbN thin-film nanowire detectors for time-of-flight mass spectrometry". *Physica C: Superconductivity*, **468**, 2001 – 2003, 2008.

[12] G. N. Gol'tsman, O. Okunev, G. Chulkova, A. Lipatov, A. Semenov, K. Smirnov, B. Voronov, A. Dzardanov, C. Williams, and R. Sobolewski. "Picosecond superconducting single-photon optical detector". *Applied Physics Letters*, **79**, 705–707, 2001.

[13] J. Yang, A. Kerman, E. Dauler, B. Cord, V. Anant, R. Molnar, and K. Berggren. "Suppressed Critical Current in Superconducting Nanowire Single-Photon Detectors With High Fill-Factors". *Applied Superconductivity, IEEE Transactions on*, **19**, 318 –322, june 2009.

[14] X. Hu, E. A. Dauler, R. J. Molnar, and K. K. Berggren. "Superconducting nanowire single-photon detectors integrated with optical nano-antennae". *Opt. Express*, **19**, 17–31, Jan 2011.

[15] T. May, J. Toussaint, R. Gruner, M. Schubert, H.-G. Meyer, B. Dietzek, J. Popp, M. Hofherr, K. Il'in, D. Henrich, M. Arndt, and M. Siegel. "Superconducting nanowire single-photon detectors for picosecond time resolved spectroscopic applications". volume 8427, p. 84274B. SPIE, 2012.

[16] X. Hu, T. Zhong, J. E. White, E. A. Dauler, F. Najafi, C. H. Herder, F. N. C. Wong, and K. K. Berggren. "Fiber-coupled nanowire photon counter at 1550 nm with 24% system detection efficiency". *Opt. Lett.*, **34**, 3607–3609, Dec 2009.

[17] E. A. Dauler, D. Rosenberg, A. J. Kerman, M. E. Grein, R. J. Molnar, J. Yoon, and J. D. Moores. "High-detection-efficiency superconducting nanowire single photon detectors with multi-mode optical coupling". In *2012 Applied Superconductivity Conference, Portland, Oregon*, 2012.

[18] G. Bachar, I. Baskin, O. Shtempeluck, and E. Buks. "On-fiber superconducting nanowire single photon detector". In *2012 Applied Superconductivity Conference, Portland, Oregon*, 2012.

[19] A. Semenov, B. Günther, U. Böttger, H.-W. Hübers, H. Bartolf, A. Engel, A. Schilling, K. Ilin, M. Siegel, R. Schneider, D. Gerthsen, and N. A. Gippius. "Optical and transport

properties of ultrathin NbN films and nanostructures". *Phys. Rev. B*, **80**, 054510, Aug 2009.

[20] M. G. Tanner, C. M. Natarajan, V. K. Pottapenjara, J. A. O?Connor, R. J. Warburton, R. H. Hadfield, B. Baek, S. Nam, S. N. Dorenbos, E. B. Ureña, T. Zijlstra, T. M. Klapwijk, and V. Zwiller. "Enhanced telecom wavelength single-photon detection with NbTiN superconducting nanowires on oxidized silicon". *Applied Physics Letters*, **96**, 221109, 2010.

[21] A. Gaggero, S. J. Nejad, F. Marsili, F. Mattioli, R. Leoni, D. Bitauld, D. Sahin, G. J. Hamhuis, R. Notzel, R. Sanjines, and A. Fiore. "Nanowire superconducting single-photon detectors on GaAs for integrated quantum photonic applications". *Applied Physics Letters*, **97**, 151108, 2010.

[22] K. M. Rosfjord, J. K. W. Yang, E. A. Dauler, A. J. Kerman, V. Anant, B. M. Voronov, G. N. Gol'tsman, and K. K. Berggren. "Nanowire single-photon detector with an integrated optical cavity and anti-reflection coating". *Opt. Express*, **14**, 527–534, Jan 2006.

[23] B. Baek, J. A. Stern, and S. W. Nam. "Superconducting nanowire single-photon detector in an optical cavity for front-side illumination". *Applied Physics Letters*, **95**, 191110, 2009.

[24] F. Marsili, V. B. Verma, J. A. Stern, S. Harrington, B. Baek, A. E. Lita, T. Gerrits, R. P. Mirin, and S. Nam. "Single-photon detectors based on tungsten silicide superconducting nanowires". In *2012 Applied Superconductivity Conference, Portland, Oregon*, 2012.

[25] J. P. Sprengers, A. Gaggero, D. Sahin, S. Jahanmirinejad, G. Frucci, F. Mattioli, R. Leoni, J. Beetz, M. Lermer, M. Kamp, S. Hofling, R. Sanjines, and A. Fiore. "Waveguide superconducting single-photon detectors for integrated quantum photonic circuits". *Applied Physics Letters*, **99**, 181110, 2011.

[26] W. Pernice, C. Schuck, O. Minaeva, M. Li, G. N. Goltsman, A. V. Sergienko, and H. X. Tang. "High Speed and High Efficiency Travelling Wave Single-Photon Detectors Embedded in Nanophotonic Circuits". *arXiv:1108.5299v1*, 2011.

[27] M. Hofherr, D. Rall, K. Ilin, M. Siegel, A. Semenov, H.-W. Hubers, and N. A. Gippius. "Intrinsic detection efficiency of superconducting nanowire single-photon detectors with different thicknesses". *Journal of Applied Physics*, **108**, 014507 –014507–9, jul 2010.

[28] M. Hofherr, D. Rall, K. S. Ilin, A. Semenov, N. Gippius, H.-W. Hübers, and M. Siegel. "Superconducting nanowire single-photon detectors: Quantum efficiency vs. film thickness". *Journal of Physics: Conference Series*, **234**, 012017, 2010.

[29] Y. Korneeva, I. Florya, A. Semenov, A. Korneev, and G. Goltsman. "New Generation of Nanowire NbN Superconducting Single-Photon Detector for Mid-Infrared". *Applied Superconductivity, IEEE Transactions on*, **21**, 323 –326, june 2011.

[30] M. Ejrnaes, R. Cristiano, O. Quaranta, S. Pagano, A. Gaggero, F. Mattioli, R. Leoni, B. Voronov, and G. Gol'tsman. "A cascade switching superconducting single photon detector". *Applied Physics Letters*, **91**, 262509, 2007.

[31] A. Casaburi, M. Ejrnaes, O. Quaranta, A. Gaggero, F. Mattioli, R. Leoni, B. Voronov, G. Gol'tsman, M. Lisitskiy, E. Esposito, C. Nappi, R. Cristiano, and S. Pagano. "Experimental characterization of NbN nanowire optical detectors with parallel stripline configuration". *Journal of Physics: Conference Series*, **97**, 012265, 2008.

[32] F. Marsili, F. Najafi, C. Herder, and K. K. Berggren. "Electrothermal simulation of superconducting nanowire avalanche photodetectors". *Applied Physics Letters*, **98**, 093507, 2011.

[33] F. Marsili, F. Najafi, E. Dauler, F. Bellei, X. Hu, M. Csete, R. J. Molnar, and K. K. Berggren. "Single-Photon Detectors Based on Ultranarrow Superconducting Nanowires". *Nano Letters*, **11**, 2048–2053, 2011.

[34] F. Marsili, F. Najafi, E. Dauler, R. J. Molnar, and K. K. Berggren. "Afterpulsing and instability in superconducting nanowire avalanche photodetectors". *Applied Physics Letters*, **100**, 112601, 2012.

[35] S. N. Dorenbos, E. M. Reiger, U. Perinetti, V. Zwiller, T. Zijlstra, and T. M. Klapwijk. "Low noise superconducting single photon detectors on silicon". *Applied Physics Letters*, **93**, 131101, 2008.

[36] K. Ilin, M. Hofherr, D. Rall, M. Siegel, A. Semenov, A. Engel, K. Inderbitzin, A. Aeschbacher, and A. Schilling. "Ultra-thin TaN Films for Superconducting Nanowire Single-Photon Detectors". *Journal of Low Temperature Physics*, **167**, 809–814, 2011.

[37] A. Engel, A. Aeschbacher, K. Inderbitzin, A. Schilling, K. Il'in, M. Hofherr, M. Siegel, A. Semenov, and H.-W. Hubers. "Tantalum nitride superconducting single-photon detectors with low cut-off energy". *Applied Physics Letters*, **100**, 062601, 2012.

[38] S. N. Dorenbos, P. Forn-Diaz, T. Fuse, A. H. Verbruggen, T. Zijlstra, T. M. Klapwijk, and V. Zwiller. "Low gap superconducting single photon detectors for infrared sensitivity". *Applied Physics Letters*, **98**, 251102, 2011.

[39] B. Baek, A. E. Lita, V. Verma, and S. W. Nam. "Superconducting a-W_xSi_{1-x} nanowire single-photon detector with saturated internal quantum efficiency from visible to 1850 nm". *Applied Physics Letters*, **98**, 251105, 2011.

[40] H. Shibata, H. Takesue, T. Honjo, T. Akazaki, and Y. Tokura. "Single-photon detection using magnesium diboride superconducting nanowires". *Applied Physics Letters*, **97**, 212504, 2010.

[41] A. Semenov, A. Engel, K. Il'in, G. Gol'tsman, M. Siegel, and H.-W. Hübers. "Ultimate performance of a superconducting quantum detector". *The European Physical Journal - Applied Physics*, **21**, 171–178, 2003.

[42] A. Annunziata, D. Santavicca, J. Chudow, L. Frunzio, M. Rooks, A. Frydman, and D. Prober. "Niobium Superconducting Nanowire Single-Photon Detectors". *Applied Superconductivity, IEEE Transactions on*, **19**, 327 –331, june 2009.

[43] J. R. Clem and K. K. Berggren. "Geometry-dependent critical currents in superconducting nanocircuits". *Phys. Rev. B*, **84**, 174510, Nov 2011.

[44] H. L. Hortensius, E. F. C. Driessen, T. M. Klapwijk, K. K. Berggren, and J. R. Clem. "Critical-current reduction in thin superconducting wires due to current crowding". *Applied Physics Letters*, **100**, 182602, 2012.

[45] D. Henrich, P. Reichensperger, M. Hofherr, J. M. Meckbach, K. Il'in, M. Siegel, A. Semenov, A. Zotova, and D. Y. Vodolazov. "Geometry-induced reduction of the critical current in superconducting nanowires". *Phys. Rev. B*, **86**, 144504, Oct 2012.

[46] A. D. Semenov, G. N. Gol'tsman, and A. A. Korneev. "Quantum detection by current carrying superconducting film". *Physica C: Superconductivity*, **351**, 349 – 356, 2001.

[47] A. Semenov, A. Engel, H.-W. Hübers, K. Il'in, and M. Siegel. "Spectral cut-off in the efficiency of the resistive state formation caused by absorption of a single-photon in current-carrying superconducting nano-strips". *European Physical Journal B*, **47**, 495, 2005.

[48] A. Engel, A. Semenov, H.-W. H. andK. Il'in, and M. Siegel. *New Frontiers in Superconductivity Research*. Nova Science Publishers, 2006.

[49] P. Haas, A. Semenov, H.-W. Hubers, P. Beyer, A. Kirste, T. Schurig, K. Il'in, M. Siegel, A. Engel, and A. Smirnov. "Spectral Sensitivity and Spectral Resolution of Superconducting Single-Photon Detectors". *Applied Superconductivity, IEEE Transactions on*, **17**, 298 –301, june 2007.

[50] J. Yang, A. Kerman, E. Dauler, V. Anant, K. Rosfjord, and K. Berggren. "Modeling the Electrical and Thermal Response of Superconducting Nanowire Single-Photon Detectors". *Applied Superconductivity, IEEE Transactions on*, **17**, 581 –585, june 2007.

[51] A. Semenov, P. Haas, H.-W. Hübers, K. Ilin, M. Siegel, A. Kirste, D. Drung, T. Schurig, and A. Engel. "Intrinsic quantum efficiency and electro-thermal model of a superconducting nanowire single-photon detector". *Journal of Modern Optics*, **56**, 345–351, 2009.

[52] A. J. Kerman, E. A. Dauler, J. K. W. Yang, K. M. Rosfjord, V. Anant, K. K. Berggren, G. N. Gol'tsman, and B. M. Voronov. "Constriction-limited detection efficiency of superconducting nanowire single-photon detectors". *Applied Physics Letters*, **90**, 101110, 2007.

[53] A. Engel, A. Semenov, H. W. Hübers, K. Il'in, and M. Siegel. "Superconducting single-photon detector for the visible and infrared spectral range". *Journal of Modern Optics*, **51**, 1459–1466, 2004.

[54] Y. Noat, T. Cren, C. Brun, F. Debontridder, V. Cherkez, K. Ilin, M. Siegel, A. Semenov, H.-W. Huebers, , and D. Roditchev. "Break-up of long-range coherence due to phase fluctuations in ultrathin superconducting NbN films". *arxiv 1205.3408v1*, 2012.

[55] A. Verevkin, J. Zhang, R. Sobolewski, A. Lipatov, O. Okunev, G. Chulkova, A. Korneev, K. Smirnov, G. N. Gol'tsman, and A. Semenov. "Detection efficiency of large-active-area NbN single-photon superconducting detectors in the ultraviolet to near-infrared range". *Applied Physics Letters*, **80**, 4687–4689, 2002.

[56] H. Bartolf, A. Engel, A. Schilling, K. Il'in, M. Siegel, H.-W. Hübers, and A. Semenov. "Current-assisted thermally activated flux liberation in ultrathin nanopatterned NbN superconducting meander structures". *Phys. Rev. B*, **81**, 024502, Jan 2010.

[57] D. Y. Vodolazov. "Saddle point states in two-dimensional superconducting films biased near the depairing current". *Phys. Rev. B*, **85**, 174507, May 2012.

[58] A. Engel, A. Semenov, H.-W. Hübers, K. Il'in, and M. Siegel. "Fluctuation effects in superconducting nanostrips". *Physica C: Superconductivity*, **444**, 12 – 18, 2006.

[59] L. N. Bulaevskii, M. J. Graf, C. D. Batista, and V. G. Kogan. "Vortex-induced dissipation in narrow current-biased thin-film superconducting strips". *Phys. Rev. B*, **83**, 144526, Apr 2011.

[60] C. P. Bean and J. D. Livingston. "Surface Barrier in Type-II Superconductors". *Phys. Rev. Lett.*, **12**, 14–16, Jan 1964.

[61] J. Pearl. "Current distribution in superconducting films carrying quantized fluxoids". *Applied Physics Letters*, **5**, 65–66, 1964.

[62] F. Tafuri, J. Kirtley, D. Born, D. Stornaiuolo, P. Medaglia, P. Orgiani, G. Balestrino, and V. Kogan. "Dissipation in ultra-thin current-carrying superconducting bridges; evidence for quantum tunneling of Pearl vortices". *EPL (Europhysics Letters)*, **73**, 948, 2006.

[63] D. D. Bacon, A. T. English, S. Nakahara, F. G. Peters, H. Schreiber, W. R. Sinclair, and R. B. van Dover. "Properties of NbN thin films deposited on ambient temperature substrates". *Journal of Applied Physics*, **54**, 6509–6516, 1983.

[64] D. Hohnke, D. Schmatz, and M. Hurley. "Reactive sputter deposition: A quantitative analysis". *Thin Solid Films*, **118**, 301 – 310, 1984.

[65] M. Deen. "The effect of the deposition rate on the properties of d.c.-magnetron-sputtered niobium nitride thin films". *Thin Solid Films*, **152**, 535 – 544, 1987.

[66] Z. Wang, A. Kawakami, Y. Uzawa, and B. Komiyama. "Superconducting properties and crystal structures of single-crystal niobium nitride thin films deposited at ambient substrate temperature". *Journal of Applied Physics*, **79**, 7837–7842, 1996.

[67] K. Tanabe, H. Asano, Y. Katoh, and O. Michikami. "Ellipsometric and optical reflectivity studies of reactively sputtered NbN thin films". *Journal of Applied Physics*, **63**, 1733–1739, 1988.

[68] A. Nigro, G. Nobile, M. G. Rubino, and R. Vaglio. "Electrical resistivity of polycrystalline niobium nitride films". *Phys. Rev. B*, **37**, 3970–3972, Mar 1988.

[69] K. Ilin, R. Schneider, D. Gerthsen, A. Engel, H. Bartolf, A. Schilling, A. Semenov, H. W. Huebers, B. Freitag, and M. Siegel. "Ultra-thin NbN films on Si: crystalline and superconducting properties". *Journal of Physics: Conference Series*, **97**, 012045, 2008.

[70] R. E. de Lamaestre, P. Odier, E. Bellet-Amalric, P. Cavalier, S. Pouget, and J.-C. Villégier. "High quality ultrathin NbN layers on sapphire for superconducting single photon detectors". *Journal of Physics: Conference Series*, **97**, 012046, 2008.

[71] J. R. Gao, M. Hajenius, F. D. Tichelaar, T. M. Klapwijk, B. Voronov, E. Grishin, G. Gol'tsman, C. A. Zorman, and M. Mehregany. "Monocrystalline NbN nanofilms on a 3C-SiC/Si substrate". *Applied Physics Letters*, **91**, 062504, 2007.

[72] S. Park and T. Geballe. "Tc depression in thin Nb films". *Physica B+C*, **135**, 108 – 112, 1985.

[73] S. A. Wolf, J. J. Kennedy, and M. Nisenoff. "Properties of superconducting rf sputtered ultrathin films of Nb". *Journal of Vacuum Science and Technology*, **13**, 145–147, 1976.

[74] L. N. Cooper. "Superconductivity in the Neighborhood of Metallic Contacts". *Phys. Rev. Lett.*, **6**, 689–690, Jun 1961.

[75] Y. V. Fominov and M. V. Feigel'man. "Superconductive properties of thin dirty superconductor–normal-metal bilayers". *Phys. Rev. B*, **63**, 094518, Feb 2001.

[76] L. Aslamasov and A. Larkin. "The influence of fluctuation pairing of electrons on the conductivity of normal metal". *Physics Letters A*, **26**, 238 – 239, 1968.

[77] K. Il'in, D. Rall, M. Siegel, A. Engel, A. Schilling, A. Semenov, and H.-W. Huebers. "Influence of thickness, width and temperature on critical current density of Nb thin film structures". *Physica C: Superconductivity*, **470**, 953 – 956, 2010.

[78] W. J. Skocpol, M. R. Beasley, and M. Tinkham. "Self-heating hotspots in superconducting thin-film microbridges". *Journal of Applied Physics*, **45**, 4054–4066, 1974.

[79] K. Ylli. "Untersuchung supraleitender Dünnfilme in einem automatisierten Messaufbau". Bachelor thesis, Karlsruhe Insitute of Technology, 2010.

[80] L. Ponta, A. Carbone, and M. Gilli. "Resistive transition in disordered superconductors with varying intergrain coupling". *Superconductor Science and Technology*, **24**, 015006, 2011.

[81] N. R. Werthamer, E. Helfand, and P. C. Hohenberg. "Temperature and Purity Dependence of the Superconducting Critical Field, H_{c2}. III. Electron Spin and Spin-Orbit Effects". *Phys. Rev.*, **147**, 295–302, Jul 1966.

[82] K. E. Kornelsen, M. Dressel, J. E. Eldridge, M. J. Brett, and K. L. Westra. "Far-infrared optical absorption and reflectivity of a superconducting NbN film". *Phys. Rev. B*, **44**, 11882–11887, Dec 1991.

[83] M. Tazawa, X. Gerbaux, J. Villègier, and A. Hadni. "A Far IR Study of NbN in the Normal and Superconductive State". *International Journal of Infrared and Millimeter Waves*, **19**, 1155–1173, 1998.

[84] D. Karecki, R. E. Peña, and S. Perkowitz. "Far-infrared transmission of superconducting homogeneous NbN films: Scattering time effects". *Phys. Rev. B*, **25**, 1565–1571, Feb 1982.

[85] D. Henrich, S. Dörner, M. Hofherr, K. Il'in, A. Semenov, E. Heintze, M. Scheffler, M. Dressel, and M. Siegel. "Broadening of hot-spot response spectrum of superconducting NbN nanowire single-photon detector with reduced nitrogen content". *Journal of Applied Physics*, **112**, 074511, 2012.

[86] M. Scheffler, J. P. Ostertag, and M. Dressel. "Fabry-Perot resonances in birefringent YAlO3 analyzed at terahertz frequencies". *Opt. Lett.*, **34**, 3520–3522, Nov 2009.

[87] D. C. Mattis and J. Bardeen. "Theory of the Anomalous Skin Effect in Normal and Superconducting Metals". *Phys. Rev.*, **111**, 412–417, Jul 1958.

[88] J. Toussaint. "Untersuchung der optischen Eigenschaften von NbN-Supraleiterschichten für die Detektoranwendung". Studienarbeit, Karlsruhe Insitute of Technology, 2010.

[89] L. R. Testardi. "Destruction of Superconductivity by Laser Light". *Phys. Rev. B*, **4**, 2189–2196, Oct 1971.

[90] M. Danerud, D. Winkler, M. Lindgren, M. Zorin, V. Trifonov, B. S. Karasik, G. N. Gol'tsman, and E. M. Gershenzon. "Nonequilibrium and bolometric photoresponse in patterned $YBa_2Cu_3O_{7-\delta}$ thin films". *Journal of Applied Physics*, **76**, 1902–1909, 1994.

[91] A. V. Sergeev, A. D. Semenov, P. Kouminov, V. Trifonov, I. G. Goghidze, B. S. Karasik, G. N. Gol'tsman, and E. M. Gershenzon. "Transparency of a $YBa_2Cu_3O_7$-film/substrate interface for thermal phonons measured by means of voltage response to radiation". *Phys. Rev. B*, **49**, 9091–9096, Apr 1994.

[92] A. D. Semenov, R. S. Nebosis, Y. P. Gousev, M. A. Heusinger, and K. F. Renk. "Analysis of the nonequilibrium photoresponse of superconducting films to pulsed radiation by use of a two-temperature model". *Phys. Rev. B*, **52**, 581–590, Jul 1995.

[93] M. Lindgren, M. Currie, C. Williams, T. Y. Hsiang, P. M. Fauchet, R. Sobolewski, S. H. Moffat, R. A. Hughes, J. S. Preston, and F. A. Hegmann. "Intrinsic picosecond response times of Y–Ba–Cu–O superconducting photodetectors". *Applied Physics Letters*, **74**, 853–855, 1999.

[94] J. W. Kooi, J. J. A. Baselmans, M. Hajenius, J. R. Gao, T. M. Klapwijk, P. Dieleman, A. Baryshev, and G. de Lange. "IF impedance and mixer gain of NbN hot electron bolometers". *Journal of Applied Physics*, **101**, 044511, 2007.

[95] S. Cherednichenko, V. Drakinskiy, J. Baubert, J.-M. Krieg, B. Voronov, G. Gol'tsman, and V. Desmaris. "Gain bandwidth of NbN hot-electron bolometer terahertz mixers on 1.5 mu m Si_3N_4/SiO_2 membranes". *Journal of Applied Physics*, **101**, 124508, 2007.

[96] K. S. Il'in and M. Siegel. "Microwave mixing in microbridges made from $YBa_2Cu_3O_{7-x}$ thin films". *Journal of Applied Physics*, **92**, 361–369, 2002.

[97] A. D. Semenov, K. Ilin, M. Siegel, A. Smirnov, S. Pavlov, H. Richter, and H.-W. Hübers. "Evidence of non-bolometric mixing in the bandwidth of a hot-electron bolometer". *Superconductor Science and Technology*, **19**, 1051, 2006.

[98] M. L. Kaganov, I. M. Lifshitz, and L. V. Tanatarov. "". *Sov. Phys JETP*, **4**, 173, 1957.

[99] N. Perrin and C. Vanneste. "Response of superconducting films to a periodic optical irradiation". *Phys. Rev. B*, **28**, 5150–5159, Nov 1983.

[100] P. Thoma, A. Scheuring, M. Hofherr, S. Wunsch, K. Il'in, N. Smale, V. Judin, N. Hiller, A.-S. Muller, A. Semenov, H.-W. Hubers, and M. Siegel. "Real-time measurement of

picosecond THz pulses by an ultra-fast $YBa_2Cu_3O_{7-d}$ detection system". *Applied Physics Letters*, **101**, 142601, 2012.

[101] K. S. Il'in, M. Lindgren, M. Currie, A. D. Semenov, G. N. Gol'tsman, R. Sobolewski, S. I. Cherednichenko, and E. M. Gershenzon. "Picosecond hot-electron energy relaxation in NbN superconducting photodetectors". *Applied Physics Letters*, **76**, 2752–2754, 2000.

[102] A. D. Semenov, G. N. Gol'tsman, and R. Sobolewski. "Hot-electron effect in superconductors and its applications for radiation sensors". *Superconductor Science and Technology*, **15**, R1, 2002.

[103] S. Wuensch, T. Ortlepp, E. Crocoll, F. Uhlmann, and M. Siegel. "Cryogenic Semiconductor Amplifier for RSFQ-Circuits With High Data Rates at 4.2 K". *Applied Superconductivity, IEEE Transactions on*, **19**, 574 –579, june 2009.

[104] M. Lindgren, M. Currie, C. Williams, T. Hsiang, P. Fauchet, R. Sobolewski, S. Moffat, R. Hughes, J. Preston, and F. Hegmann. "Ultrafast photoresponse in microbridges and pulse propagation in transmission lines made from high-Tc superconducting Y-Ba-Cu-O thin films". *Selected Topics in Quantum Electronics, IEEE Journal of*, **2**, 668 –678, sep 1996.

[105] D. Rall, P. Probst, M. Hofherr, S. Wünsch, K. Il'in, U. Lemmer, and M. Siegel. "Energy relaxation time in NbN and YBCO thin films under optical irradiation". *Journal of Physics: Conference Series*, **234**, 042029, 2010.

[106] E. Gershenzon, M. Gershenzon, G. Gol'tsman, A. Semenov, and A. Sergeev. "Wideband highspeed Nb and YBaCuO detectors". *Magnetics, IEEE Transactions on*, **27**, 2836 –2839, mar 1991.

[107] D. Dew-Hughes. "The critical current of superconductors: an historical review". *Low Temperature Physics*, **27**, 713–722, 2001.

[108] M. Tinkham. *Introduction to superconductivity*. McGraw-Hill, NY, 1996.

[109] M. Y. Kupriyanov and V. F. Lukichev. "Temperature dependence of pair-breaking current in superconductors". *Sov. J. Low Temp. Phys.*, **6(4)**, 210–14, 1980.

[110] K. K. Likharev. "Superconducting weak links". *Rev. Mod. Phys.*, **51**, 101–159, Jan 1979.

[111] G. Stan, S. B. Field, and J. M. Martinis. "Critical Field for Complete Vortex Expulsion from Narrow Superconducting Strips". *Phys. Rev. Lett.*, **92**, 097003, Mar 2004.

[112] E. Zeldov, J. R. Clem, M. McElfresh, and M. Darwin. "Magnetization and transport currents in thin superconducting films". *Phys. Rev. B*, **49**, 9802–9822, Apr 1994.

[113] M. Y. Kupriyanov and K. K. Likharev. "Effect of an edge barrier on the critical current of superconducting films". *Sov. Phys. Solid State*, **16**, 1835–7, April 1975.

[114] B. L. T. Plourde, D. J. Van Harlingen, D. Y. Vodolazov, R. Besseling, M. B. S. Hesselberth, and P. H. Kes. "Influence of edge barriers on vortex dynamics in thin weak-pinning superconducting strips". *Phys. Rev. B*, **64**, 014503, Jun 2001.

[115] G. Maksimova. "Mixed state and critical current in narrow semiconducting films". *Physics of the Solid State*, **40**, 1607–1610, 1998.

[116] O. V. Dobrovolskiy and M. Huth. "Crossover from dirty to clean superconducting limit in dc magnetron-sputtered thin Nb films". *Thin Solid Films*, **520**, 5985 – 5990, 2012.

[117] A. Y. Rusanov, M. B. S. Hesselberth, and J. Aarts. "Depairing currents in superconducting films of Nb and amorphous MoGe". *Phys. Rev. B*, **70**, 024510, Jul 2004.

[118] Y. Kamihara, T. Watanabe, M. Hirano, and H. Hosono. "Iron-Based Layered Superconductor $La(O_{1-x}F_x)FeAs$ (x = 0.05 0.12) with Tc = 26 K". *Journal of the American Chemical Society*, **130**, 3296–3297, 2008.

[119] M. Rotter, M. Tegel, and D. Johrendt. "Superconductivity at 38 K in the Iron Arsenide $(Ba_{1-x}K_x)Fe_2As_2$". *Phys. Rev. Lett.*, **101**, 107006, Sep 2008.

[120] X. Wang, Q. Liu, Y. Lv, W. Gao, L. Yang, R. Yu, F. Li, and C. Jin. "The superconductivity at 18 K in LiFeAs system". *Solid State Communications*, **148**, 538 – 540, 2008.

[121] F.-C. Hsu, J.-Y. Luo, K.-W. Yeh, T.-K. Chen, T.-W. Huang, P. M. Wu, Y.-C. Lee, Y.-L. Huang, Y.-Y. Chu, D.-C. Yan, and M.-K. Wu. "Superconductivity in the PbO-type structure α-FeSe". *Proceedings of the National Academy of Sciences*, **105**, 14262–14264, 2008.

[122] K. Iida, J. Hanisch, R. Huhne, F. Kurth, M. Kidszun, S. Haindl, J. Werner, L. Schultz, and B. Holzapfel. "Strong T[sub c] dependence for strained epitaxial $Ba(Fe_{1-x}Co_x)_2As_2$ thin films". *Applied Physics Letters*, **95**, 192501, 2009.

[123] A. S. Sefat, R. Jin, M. A. McGuire, B. C. Sales, D. J. Singh, and D. Mandrus. "Superconductivity at 22 K in Co-Doped $BaFe_2As_2$ Crystals". *Phys. Rev. Lett.*, **101**, 117004, Sep 2008.

[124] M. Putti, I. Pallecchi, E. Bellingeri, M. R. Cimberle, M. Tropeano, C. Ferdeghini, A. Palenzona, C. Tarantini, A. Yamamoto, J. Jiang, J. Jaroszynski, F. Kametani, D. Abraimov, A. Polyanskii, J. D. Weiss, E. E. Hellstrom, A. Gurevich, D. C. Larbalestier, R. Jin, B. C. Sales, A. S. Sefat, M. A. McGuire, D. Mandrus, P. Cheng, Y. Jia, H. H.

Wen, S. Lee, and C. B. Eom. "New Fe-based superconductors: properties relevant for applications". *Superconductor Science and Technology*, **23**, 034003, 2010.

[125] S. A. Baily, Y. Kohama, H. Hiramatsu, B. Maiorov, F. F. Balakirev, M. Hirano, and H. Hosono. "Pseudoisotropic Upper Critical Field in Cobalt-Doped $SrFe_2As_2$ Epitaxial Films". *Phys. Rev. Lett.*, **102**, 117004, Mar 2009.

[126] T. Katase, H. Hiramatsu, H. Yanagi, T. Kamiya, M. Hirano, and H. Hosono. "Atomically-flat, chemically-stable, superconducting epitaxial thin film of iron-based superconductor, cobalt-doped". *Solid State Communications*, **149**, 2121 – 2124, 2009.

[127] K. Iida, S. Haindl, T. Thersleff, J. Hanisch, F. Kurth, M. Kidszun, R. Huhne, I. Monch, L. Schultz, B. Holzapfel, and R. Heller. "Influence of Fe buffer thickness on the crystalline quality and the transport properties of $Fe/Ba(Fe_{1-x}Co_x)_2As_2$ bilayers". *Applied Physics Letters*, **97**, 172507, 2010.

[128] L. Luan, O. M. Auslaender, T. M. Lippman, C. W. Hicks, B. Kalisky, J.-H. Chu, J. G. Analytis, I. R. Fisher, J. R. Kirtley, and K. A. Moler. "Local measurement of the penetration depth in the pnictide superconductor $Ba(Fe_{0.95}Co_{0.05})_2As_2$". *Phys. Rev. B*, **81**, 100501, Mar 2010.

[129] A. Yamamoto, J. Jaroszynski, C. Tarantini, L. Balicas, J. Jiang, A. Gurevich, D. C. Larbalestier, R. Jin, A. S. Sefat, M. A. McGuire, B. C. Sales, D. K. Christen, and D. Mandrus. "Small anisotropy, weak thermal fluctuations, and high field superconductivity in Co-doped iron pnictide $Ba(Fe_{1-x}Co_x)_2As_2$". *Applied Physics Letters*, **94**, 062511, 2009.

[130] R. Prozorov, N. Ni, M. A. Tanatar, V. G. Kogan, R. T. Gordon, C. Martin, E. C. Blomberg, P. Prommapan, J. Q. Yan, S. L. Bud'ko, and P. C. Canfield. "Vortex phase diagram of $Ba(Fe_{0.93}Co_{0.07})_2As_2$ single crystals". *Phys. Rev. B*, **78**, 224506, Dec 2008.

[131] K. Iida, J. Hänisch, T. Thersleff, F. Kurth, M. Kidszun, S. Haindl, R. Hühne, L. Schultz, and B. Holzapfel. "Scaling behavior of the critical current in clean epitaxial $Ba(Fe_{1-x}Co_x)_2As_2$ thin films". *Phys. Rev. B*, **81**, 100507, Mar 2010.

[132] T. Katase, Y. Ishimaru, A. Tsukamoto, H. Hiramatsu, T. Kamiya, K. Tanabe, and H. Hosono. "Josephson junction in cobalt-doped $BaFe_2As_2$ epitaxial thin films on $(La,Sr)(Al,Ta)O_3$ bicrystal substrates". *Applied Physics Letters*, **96**, 142507, 2010.

[133] C. Tarantini, S. Lee, Y. Zhang, J. Jiang, C. W. Bark, J. D. Weiss, A. Polyanskii, C. T. Nelson, H. W. Jang, C. M. Folkman, S. H. Baek, X. Q. Pan, A. Gurevich, E. E. Hellstrom, C. B. Eom, and D. C. Larbalestier. "Strong vortex pinning in Co-doped $BaFe_2As_2$ single crystal thin films". *Applied Physics Letters*, **96**, 142510, 2010.

[134] S. Lee, J. Jiang, Y. Zhang, C. W. Bark, J. D. Weiss, C. Tarantini, C. T. Nelson, H. W. Jang, C. M. Folkman, S. H. Baek, A. Polyanskii, D. Abraimov, A. Yamamoto, J. W. Park, X. Q. Pan, E. E. Hellstrom, D. C. Larbalestier, and C. B. Eom. "Template engineering of Co-doped $BaFe_2As_2$ single-crystal thin films". *Nature Materials*, **9**, 397–402, 2010.

[135] D. Rall, K. Il'in, K. Iida, S. Haindl, F. Kurth, T. Thersleff, L. Schultz, B. Holzapfel, and M. Siegel. "Critical current densities in ultrathin Ba(Fe,Co)$_2$As$_2$ microbridges". *Phys. Rev. B*, **83**, 134514, Apr 2011.

[136] D. Rall, L. Rehm, K. Ilin, M. Siegel, K. Iida, S. Haindl, F. Kurth, B. Holzapfel, L. Schultz, J. Yong, and T. Lemberger. "Penetration and de-pinning of vortices in sub-micrometer Ba(Fe,Co)2As2 thin film bridges". *Physica C: Superconductivity*, **479**, 164 – 166, 2012.

[137] S. Schmidt, S. Döring, F. Schmidl, V. Grosse, P. Seidel, K. Iida, F. Kurth, S. Haindl, I. Mönch, and B. Holzapfel. "BaFe1.8Co0.2As2 thin film hybrid Josephson junctions". *Applied Physics Letters*, **97**, 172504, 2010.

[138] L. Rehm, D. Henrich, M. Hofherr, S. Wuensch, P. Thoma, A. Scheuring, K. Il'in, M. Siegel, S. Haindl, K. Iida, F. Kurth, B. Holzapfel, and L. Schultz. "Infrared Photo-Response of Fe-Shunted Ba-122 Thin Film Microstructures". *Applied Superconductivity, IEEE Transactions on*, **23**, 7501105–7501105, 2013.

[139] L. Rehm. "Detektoren aus Ferroarsenid-Dünnschichten". Diplomarbeit, Karlsruhe Insitute of Technology, 2012.

[140] J. Yong, S. Lee, J. Jiang, C. W. Bark, J. D. Weiss, E. E. Hellstrom, D. C. Larbalestier, C. B. Eom, and T. R. Lemberger. "Superfluid density measurements of Ba(Co$_x$Fe$_{1-x}$)$_2$As$_2$ films near optimal doping". *Phys. Rev. B*, **83**, 104510, Mar 2011.

[141] G. M. Maksimova, N. V. Zhelezina, and I. L. Maksimov. "Critical current and negative magnetoresistance of superconducting film with edge barrier". *EPL (Europhysics Letters)*, **53**, 639, 2001.

[142] K. Il'in, M. Siegel, A. Engel, H. Bartolf, A. Schilling, A. Semenov, and H.-W. Huebers. "Current-Induced Critical State in NbN Thin-Film Structures". *Journal of Low Temperature Physics*, **151**, 585–590, 2008.

[143] A. Engel, H. Bartolf, A. Schilling, K. Il'in, M. Siegel, A. Semenov, and H.-W. Hübers. "Temperature- and field-dependence of critical currents in NbN microbridges". *Journal of Physics: Conference Series*, **97**, 012152, 2008.

[144] K. S. Ilin, D. Rall, M. Siegel, and A. Semenov. "Critical current density in thin superconducting TaN film structures". *Physica C: Superconductivity*, **479**, 176 – 178, 2012.

[145] F. B. Hagedorn and P. M. Hall. "Right-Angle Bends in Thin Strip Conductors". *Journal of Applied Physics*, **34**, 128–133, 1963.

[146] J. R. Clem, Y. Mawatari, G. R. Berdiyorov, and F. M. Peeters. "Predicted field-dependent increase of critical currents in asymmetric superconducting nanocircuits". *Phys. Rev. B*, **85**, 144511, Apr 2012.

[147] P. Reichensperger. "Design eines SNSPD mit optimierter Stromdichteverteilung". Bachelor thesis, Karlsruhe Insitute of Technology, 2011.

[148] Y. Luck. "Geometrieabhängigkeit von j_C in Dünnschicht-Niobbrücken". Bachelor thesis, Karlsruhe Institute of Technology, 2012.

[149] A. F. Mayadas, R. B. Laibowitz, and J. J. Cuomo. "Electrical Characteristics of rf-Sputtered Single-Crystal Niobium Films". *Journal of Applied Physics*, **43**, 1287–1289, 1972.

[150] H. Bartolf, K. Inderbitzin, L. B. Gómez, A. Engel, A. Schilling, H.-W. Hübers, A. Semenov, K. Il'in, and M. Siegel. "Symbiotic Optimization of the Nanolithography and RF-Plasma Etching for Fabricating High-Quality Light-Sensitive Superconductors on the 50 nm Scale". *arXiv:1011.4676*.

[151] M. Hofherr. "Entwicklung eines Messaufbaus zur optischen Charakterisierung sehr dünner NbN Schichten". Diplomarbeit, Karlsruhe Institute of Technology, 2008.

[152] S. Dörner. "Erhöhung der spektralen Bandbreite von NbN-SNSPDs". Bachelor thesis, Karlsruhe Insitute of Technology, 2012.

[153] S. N. Dorenbos, E. M. Reiger, N. Akopian, U. Perinetti, V. Zwiller, T. Zijlstra, and T. M. Klapwijk. "Superconducting single photon detectors with minimized polarization dependence". *Applied Physics Letters*, **93**, 161102, 2008.

[154] V. Anant, A. J. Kerman, E. A. Dauler, J. K. W. Yang, K. M. Rosfjord, and K. K. Berggren. "Optical properties of superconducting nanowire single-photon detectors". *Opt. Express*, **16**, 10750–10761, Jul 2008.

[155] T. Yamashita, S. Miki, K. Makise, W. Qiu, H. Terai, M. Fujiwara, M. Sasaki, and Z. Wang. "Origin of intrinsic dark count in superconducting nanowire single-photon detectors". *Applied Physics Letters*, **99**, 161105, 2011.

[156] T. Yamashita, S. Miki, W. Qiu, M. Fujiwara, M. Sasaki, and Z. Wang. "Temperature Dependent Performances of Superconducting Nanowire Single-Photon Detectors in an Ultralow-Temperature Region". *Applied Physics Express*, **3**, 102502, 2010.

[157] M. Hofherr, D. Rall, K. Il'in, A. Semenov, H.-W. Hübers, H-Wbers, and M. Siegel. "Dark Count Suppression in Superconducting Nanowire Single Photon Detectors". *Journal of Low Temperature Physics*, **167, 5**, 822–826, 2012.

[158] A. Stockhausen, K. Il'in, M. Siegel, U. Södervall, P. Jedrasik, A. Semenov, and H.-W. Hübers. "Adjustment of self-heating in long superconducting thin film NbN micro-bridges". *Superconductor Science and Technology*, **25**, 035012, 2012.

[159] J. Kitaygorsky, J. Zhang, A. Verevkin, A. Sergeev, A. Korneev, V. Matvienko, P. Kouminov, K. Smirnov, B. Voronov, G. Gol'tsman, and R. Sobolewski. "Origin of dark counts in nanostructured NbN single-photon detectors". *Applied Superconductivity, IEEE Transactions on*, **15**, 545 – 548, june 2005.

[160] J. Kitaygorsky, I. Komissarov, A. Jukna, D. Pan, O. Minaeva, N. Kaurova, A. Divochiy, A. Korneev, M. Tarkhov, B. Voronov, I. Milostnaya, G. Gol'tsman, and R. Sobolewski. "Dark Counts in Nanostructured NbN Superconducting Single-Photon Detectors and Bridges". *Applied Superconductivity, IEEE Transactions on*, **17**, 275 –278, june 2007

[161] A. Engel, K. Inderbitzin, A. Schilling, R. Lusche, A. Semenov, H.-W. Hübers, D. Henrich, M. Hofherr, K. Il'in, and M. Siegel. "Temperature-Dependence of Detection Efficiency in NbN and TaN SNSPD". *Applied Superconductivity, IEEE Transactions on*, **23**, 2300505–2300505, 2013.

[162] D. Henrich, L. Rehm, S. Dorner, M. Hofherr, K. Il'in, A. Semenov, and M. Siegel. "Detection Efficiency of a Spiral-Nanowire Superconducting Single-Photon Detector". *Applied Superconductivity, IEEE Transactions on*, **23**, 2200405–2200405, 2013.

[163] M. K. Akhlaghi, H. Atikian, A. Eftekharian, M. Loncar, and A. H. Majedi. "Reduced dark counts in optimized geometries for superconducting nanowire single photon detectors". *Opt. Express*, **20**, 23610–23616, Oct 2012.

Karlsruher Schriftenreihe zur Supraleitung
(ISSN 1869-1765)

Herausgeber: Prof. Dr.-Ing. M. Noe, Prof. Dr. rer. nat. M. Siegel

Die Bände sind unter www.ksp.kit.edu als PDF frei verfügbar oder als Druckausgabe bestellbar.

Band 001
Christian Schacherer
Theoretische und experimentelle Untersuchungen zur Entwicklung supraleitender resistiver Strombegrenzer. 2009
ISBN 978-3-86644-412-6

Band 002
Alexander Winkler
Transient behaviour of ITER poloidal field coils. 2011
ISBN 978-3-86644-595-6

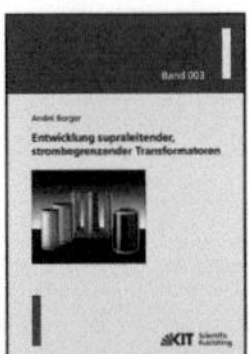

Band 003
André Berger
Entwicklung supraleitender, strombegrenzender Transformatoren. 2011
ISBN 978-3-86644-637-3

Band 004
Christoph Kaiser
High quality Nb/Al-AlOx/Nb Josephson junctions. Technological development and macroscopic quantum experiments. 2011
ISBN 978-3-86644-651-9

Band 005
Gerd Hammer
Untersuchung der Eigenschaften von planaren Mikrowellen-resonatoren für Kinetic-Inductance Detektoren bei 4,2 K. 2011
ISBN 978-3-86644-715-8

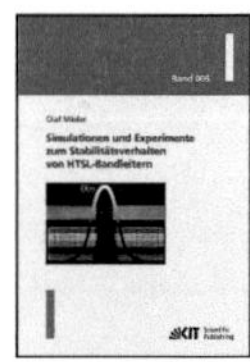

Band 006
Olaf Mäder
**Simulationen und Experimente zum Stabilitätsverhalten
von HTSL-Bandleitern.** 2012
ISBN 978-3-86644-868-1

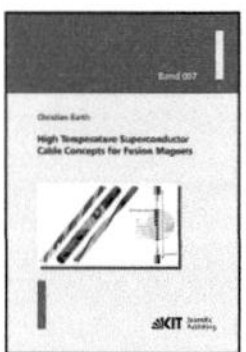

Band 007
Christian Barth
**High Temperature Superconductor Cable Concepts for
Fusion Magnets.** 2013
ISBN 978-3-7315-0065-0

Band 008
Axel Stockhausen
Optimization of Hot-Electron Bolometers for THz Radiation. 2013
ISBN 978-3-7315-0066-7

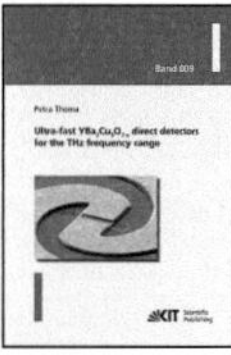

Band 009
Petra Thoma
**Ultra-fast $YBa_2Cu_3O_{7-x}$ direct detectors for the THz frequency
range.** 2013
ISBN 978-3-7315-0070-4

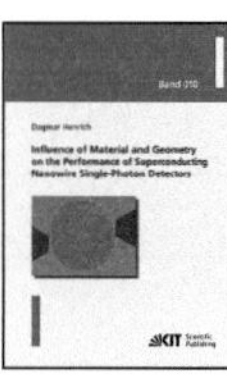

Band 010
Dagmar Henrich
**Influence of Material and Geometry on the Performance
of Superconducting Nanowire Single-Photon Detectors.** 2013
ISBN 978-3-7315-0092-6